石墨烯技术
前沿丛书

氧化石墨烯制备与应用

朱彦武　马宇飞　唐润理　主编

化学工业出版社
·北京·

《氧化石墨烯制备与应用》对氧化石墨烯相关研究成果进行了系统总结和归纳。本书前半部分从氧化石墨烯的化学基础开始，讲述氧化石墨烯的制备、结构、表征（第1章）和还原修饰、改性（第2章与第3章）以及如何将氧化石墨烯进行组装（第4章）。书中后半部分展示了氧化石墨烯在无机复合材料（第5章）、能量存储（第6章）、生物医学（第7章）、水处理（第8章）和光电传感（第9章）等领域中的典型应用及相关影响因素。

本书希望为初步进入该领域的研究生和工程师提供一个关于氧化石墨烯较为系统和全面的介绍。

图书在版编目（CIP）数据

氧化石墨烯制备与应用/朱彦武，马宇飞，唐润理主编．—北京：化学工业出版社，2019.6
（石墨烯技术前沿丛书）
ISBN 978-7-122-34241-6

Ⅰ.①氧⋯　Ⅱ.①朱⋯ ②马⋯ ③唐⋯　Ⅲ.①石墨-纳米材料-研究　Ⅳ.①TB383

中国版本图书馆CIP数据核字（2019）第060634号

责任编辑：陶艳玲
责任校对：刘　颖　　　　　　　　　　装帧设计：刘丽华

出版发行：化学工业出版社（北京市东城区青年湖南街13号　邮政编码100011）
印　　刷：北京京华铭诚工贸有限公司
装　　订：三河市振勇印装有限公司
710mm×1000mm　1/16　印张12½　字数242千字　2019年9月北京第1版第1次印刷

购书咨询：010-64518888　　售后服务：010-64518899
网　　址：http://www.cip.com.cn
凡购买本书，如有缺损质量问题，本社销售中心负责调换。

定　　价：78.00元　　　　　　　　　　　　　　版权所有　违者必究

Preface

Graphene oxide (GO), an individual layer from graphite oxide (GO) has been recognized as a versatile and valuable material only recently, despite studies on graphite oxide that extend back to the 1850s. "Expandable" graphite, typically prepared by intercalating graphite with oxidants such as sulfuric acid, is a valuable material such as for its strong expansion upon heating. Graphene can be considered a relative of graphene oxide (GO) or vice versa. Thousands of publications on fundamental research and applications have appeared in about the past 12-13 years, and this is in part because GO can be fairly readily prepared at the individual layer level. For example, while progress is being made, there is no method currently to exfoliate graphite to predominantly individual layers of graphene. GO has been explored for its potential use in electronics, energy storage, biomedicine, environmental protection, membranes, fabrics, compounds, fibers, and others. After partial removal of oxygen from GO, the carbon "backbone" recovers some features of graphene. The ability to obtain a high yield of single layer GO has thus motivated its conversion to "near-graphene" like individual layers, and the cost factor has motivated industry to be intensely interested in this approach.

It seems to be the case that thousands of companies primarily in Europe, East Asia, and North America, have been started or adopted GO-based products. GO-based products are achieving, it seems, commercial success, and (annual) billion-dollar level markets for individual products are likely, soon.

It is thus timely to have a well-structured and systematic overview of research done on GO, for scientists and engineers (including students as well as professionals) with an interest. The authors have significant experience both in fundamental R&D *and* in industrialization of GO materials.

This book starts by describing the basic chemistry related to graphene oxide including its preparation, characterization, and chemical modification, and then goes on to describe how GO is or has been assembled into complex structures or systems. Applications of interest to industry, such as composites with polymeric compounds, energy, biomedicine, water treatment, and photoelectric sensing, are covered next.

This comprehensive book will be of great use for Chinese readers as the readers can readily appreciate and learn more deeply about the major achievements in R&D

of GO materials and trends. The 9 chapters in this book provide a "state-of-the-art" overview of R&D, and even now mature or nearly mature, applications. The chapters are well organized and each builds on the prior chapter, but individual chapters can also be read independently depending on the reader's interest.

Rodney S. Ruoff
Director, IBS Center for Multidimensional Carbon Materials
Distinguished Professor, Ulsan National Institute of Science and Technology

前言

PREFACE

尽管对石墨氧化物的研究可以追溯到大约150年以前，但是直到近些年人们才意识到，石墨氧化物是一类非常有用的材料：轻微氧化后得到可膨胀石墨，可进一步加工成诸多石墨制品；充分氧化得到氧化石墨，解理后可以制备氧化石墨烯或者经还原、改性后得到石墨烯及各种衍生物。在研究过程中，人们也逐渐认识到，对石墨进行氧化，不单可以使其层间膨胀从而便于解理，更是对石墨这一天然碳晶体材料进行结构和化学调控的重要手段；这一路径不但可用于得到各种石墨衍生物，还可成为化学家、材料学家等研究碳基结构和表面的有用"平台"。

随着石墨烯等二维材料引起巨大关注，过去十余年来针对氧化石墨的相关研究出现井喷，发表的论文和专利数以万计。石墨氧化法被认为是规模化制备石墨烯的主要方法之一，对于氧化石墨相关石墨烯的研究引起了工业界的密切关注。世界主要经济体都竞相制定政策，规划大量投入进行研究和开发，催生了众多从事石墨烯相关业务的企业，一些产品成功实现应用，规模化市场开始出现。然而，由于对于氧化石墨相关的研究涉及众多应用领域，在不同研究报道中因原料选择、制备方法、产品特性、评估方法、应用场景的不同，也会出现一些有争议的结果，甚至可能会对非专业人士和行业发展造成误导。

基于以上背景，本书从氧化石墨烯的化学基础开始，对氧化石墨烯相关材料的制备、表征和若干代表性应用进行了系统总结,重点在于突出能够实现特定应用的某些特性以及对这些应用效果的综述评价。本书的主要目的是为初次进入此领域的研究生和工程师们提供一个介绍,使之能对氧化石墨烯相关的基本概念、主要应用和相关技术进行全面了解。在此基础上进一步的研究和开发,还需要研究人员针对特定内容、参考相关文献进行更加仔细的阅读与分析。细心的读者大概已经看出,对这类材料的描述,在本文短短篇幅中已经出现了"石墨氧化物""氧化石墨""氧化石墨烯"等不同名称,可以看出这类材料体系中不同处理过程和特性带来的命名复杂性。由于"氧化石墨烯"一词通常用于表述氧化石墨解理后的单层薄片,并可进一步组装、复合或者还原得到多种石墨烯基材料,为方便起见,本书大部分地方以此统称从氧化石墨得到的相关石墨烯材料,仅在若干必要的地方明确加以区分。但是希望读者能够分辨上述名称的不同内涵。

本书的大部分作者兼具基础研究和产品开发经验,希望能够从更加实用的角度为氧化石墨烯类材料提供评述,并尽可能使本书为不具有相关专业背景的人员所理解。因此,虽然尽量努力保证本书的科学性,但在上述原则下难免会牺牲一点严格性。而且,关于氧化石墨烯的文献浩如烟海,且不断有新成果涌现,毫无疑问本书无法做到全部囊括。限于编者水平,书中不足之处还请读者批评指正。如有不同意见和其他建议,也欢迎共同探讨。

编者

2019 年 2 月

常用英文缩写对照表

英文全称	英文简称	中文全称
Atomic Force Microscope	AFM	原子力显微镜
Anodic Aluminum Oxide	AAO	多孔阳极氧化铝
Alkaline Fuel Cell	AFC	碱性燃料电池
Brominated Butyl Rubber	BSA	牛血清白蛋白
Butyl Bromide	BIIR	溴化丁基橡胶
Carbon Nanotubes	CNTs	碳纳米管
Chemical Vapor Deposition	CVD	化学气相沉积
Cross Polarization-Magic Angle Spinning	CP-MAS	交叉极化/魔角旋转
Doxorubicin	DOX	阿霉素
Diffuse reflectance Infrared Fourier Transform Spectroscopy	DRIFT	漫反射傅里叶红外光谱
Differential Scanning Calorimetry	DSC	差示扫描量热法
Dye Sensitized Solar Cell	DSSC	染料敏化太阳能电池
Electrochemical Synthesized Graphene Oxide	EGO	电化学合成氧化石墨烯
Electrical Double-Layer Capacitor	EDLC	双电层电容器
Enhanced Permeability and Retention Effect	EPR	滞留效应
Graphene Oxide	GO	氧化石墨烯
Graphite Intercalation Compound	GIC	石墨层间化合物
Glutathione	GSH	L-谷胱甘肽
Graphene Nanoribbons	GNR	石墨烯纳米带
Graphene Fiber Supercapacitor	GFSC	石墨烯纤维超级电容器
Graphene Fibers	GFs	石墨烯纤维
Glass Transition Temperature	T_g	玻璃化转变温度
Highly Oriented Pyrolytic Graphite	HOPG	高定向热解石墨
High Resolution Transmission Electron Microscope	HRTEM	高分辨率的透射电子显微镜
Indium Tin Oxide	ITO	氧化铟锡
Ionic Liquid	IL	离子液体
Liquid Silicone Rubber	LSR	液态硅橡胶
Magic Angle Spinning-Nuclear Magnetic Resonance	MAS-NMR	转角核磁共振
Molecular Initiating Event	MIE	起始分子反应

Methylthio Tetrazole	MTT	四甲基偶氮唑盐
Methyl Methacrylate	MMA	甲基丙烯酸甲酯
Molten-Carbonate Fuel Cell	MCFC	隔膜熔融碳酸燃料电池
N,N-Dimethyl Formamide	DMF	N,N-二甲基甲酰胺
Nuclear Magnetic Resonance	NMR	核磁共振
Natural Rubber	NR	天然橡胶
Organic Photovoltaic Cell	OPV	有机光伏电池
para-Phenylene Diamine	PPD	对苯二胺
Polyvinyl Alcohol	PVA	聚乙烯醇
Polymethyl Methacrylate	PMMA	聚甲基丙烯酸甲酯
Polystyrene	PS	聚苯乙烯
Polyvinylidene Difluoride	PVDF	聚偏二氟乙烯
Polyetherimide	PEI	聚乙烯亚胺
Polyaniline	PANI	聚苯胺
Polypropylene	PP	聚丙烯
Polyvinyl Chloride	PVC	聚氯乙烯
Polypyrrole	PPY	聚吡咯
Poly(3,4-Ethylenedioxythiophene)	PEDOT	聚乙烯二氧噻吩
Proton Exchange Membrane Fuel Cell	PEMFC	质子交换膜燃料电池
Phosphoric Acid Fuel Cell	PAFC	磷酸燃料电池
Reduced Graphene Oxide	RGO	还原氧化石墨烯
Styrene-Butadiene Rubber	SBR	丁苯橡胶
Solid State Nuclear Magnetic Resonance	SSNMR	固体核磁共振
Scanning Transmission Electron Microscope	STEM	扫描透射电子显微镜
Scanning Tunneling Microscope	STM	扫描隧道显微镜
Single-Walled Carbon Nanotube	SWCNT	单壁碳纳米管
Semi-Ordered Aggregates	SOA	半有序聚集体
Solid Oxide Fuel Cell	SOFC	固体氧化物燃料电池
Thermogravimetric Analysis	TGA	热重分析
Transparent Conductive Film	TCF	透明导电薄膜
Transmission Electron Microscope	TEM	透射电子显微镜
Triethoxysilane	TEVS	三乙氧基硅烷
Vinyl Polyester Resin	VPR	乙烯基聚酯树脂
X-ray Diffraction	XRD	X射线衍射
X-ray Photoelectron Spectroscopy	XPS	X射线光电子能谱

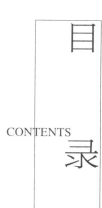

第1章 氧化石墨烯的制备及表征 /001

1.1 合成 …………………………………………… 002
1.1.1 Brodie's 法 ………………………………… 002
1.1.2 Staudenmaier's 法 …………………………… 002
1.1.3 Hummers' 法 ………………………………… 003
1.1.4 Tour's 法 …………………………………… 004
1.1.5 高铁酸钾氧化法 ……………………………… 005
1.1.6 电化学制备法 ………………………………… 006
1.1.7 其他氧化石墨烯的制备方法 ………………… 008
1.2 氧化石墨烯的结构研究 ……………………… 008
1.3 氧化石墨烯的光谱特征和结构表征方法 …… 011
1.3.1 氧化石墨烯的光谱特征 ……………………… 011
1.3.2 氧化石墨烯的形态学特征及热稳定性 ……… 013
1.4 小结 …………………………………………… 017
参考文献 …………………………………………… 018

第2章 氧化石墨的还原 /021

2.1 氧化石墨烯还原评价方法 …………………… 021
2.1.1 光学特征 ……………………………………… 022
2.1.2 导电性 ………………………………………… 023
2.1.3 碳氧比 ………………………………………… 023
2.1.4 其他技术 ……………………………………… 024
2.2 还原方法与机理 ……………………………… 024

2.2.1　热还原 ………………………………………………………… 024
　　2.2.2　光催化还原 …………………………………………………… 028
　　2.2.3　溶剂热还原 …………………………………………………… 028
　　2.2.4　电化学还原 …………………………………………………… 029
　　2.2.5　化学试剂还原 ………………………………………………… 029
　　2.2.6　植物提取物还原 ……………………………………………… 039
　　2.2.7　微生物还原 …………………………………………………… 039
　　2.2.8　蛋白质还原 …………………………………………………… 040
　　2.2.9　激素还原 ……………………………………………………… 040
　2.3　小结 …………………………………………………………………… 041
　参考文献 …………………………………………………………………… 041

第3章　氧化石墨烯的修饰和改性　/047

　3.1　氧化石墨烯的还原改性 ……………………………………………… 048
　3.2　共价修饰改性 ………………………………………………………… 048
　　3.2.1　羧基修饰 ……………………………………………………… 049
　　3.2.2　环氧基团修饰 ………………………………………………… 049
　　3.2.3　羟基修饰反应 ………………………………………………… 050
　3.3　聚合物共价修饰氧化石墨烯表面 …………………………………… 051
　　3.3.1　Graft from 接枝修饰方式 …………………………………… 051
　　3.3.2　Graft to 修饰改性氧化石墨烯表面 ………………………… 053
　3.4　非共价键修饰 ………………………………………………………… 055
　3.5　氧化石墨烯交联 ……………………………………………………… 057
　3.6　小结 …………………………………………………………………… 059
　参考文献 …………………………………………………………………… 059

第4章　氧化石墨烯组装　/063

　4.1　氧化石墨烯纤维（一维）组装及应用 ……………………………… 063
　　4.1.1　氧化石墨烯纤维的制备及性能研究 ………………………… 064
　　4.1.2　氧化石墨烯纤维的主要应用 ………………………………… 070
　4.2　氧化石墨烯薄膜结构组装及应用 …………………………………… 072
　　4.2.1　抽滤成膜法 …………………………………………………… 072
　　4.2.2　Langmuir-Blodgett（LB）组装法 …………………………… 075
　　4.2.3　逐层组装法 …………………………………………………… 076
　　4.2.4　界面诱导自组装 ……………………………………………… 076
　　4.2.5　旋转涂膜法 …………………………………………………… 077
　　4.2.6　其他方法 ……………………………………………………… 078
　4.3　氧化石墨烯三维宏观体结构组装及应用 …………………………… 079

  4.3.1 水热法 ·· 080
  4.3.2 溶剂热法 ·· 081
  4.3.3 化学还原中的组装 ·· 082
  4.3.4 三维组装中的交联方法 ·· 082
  4.3.5 模板法 ·· 084
  4.3.6 其他方法 ·· 085
 4.4 小结 ··· 086
 参考文献 ··· 086

第5章 氧化石墨烯基复合材料 /094

 5.1 氧化石墨烯/高分子复合材料体系 ····································· 094
  5.1.1 制备方法 ·· 094
  5.1.2 应用 ··· 096
 5.2 氧化石墨烯/无机非金属复合材料体系 ······························ 105
  5.2.1 力学性能 ·· 106
  5.2.2 耐久性 ·· 109
  5.2.3 其他方面 ·· 109
 5.3 小结 ··· 110
 参考文献 ··· 110

第6章 氧化石墨烯在能量存储中的应用 /119

 6.1 氧化石墨烯在锂离子电池中应用的研究进展 ···················· 119
  6.1.1 氧化石墨烯用作锂离子电池负极材料 ···················· 120
  6.1.2 氧化石墨烯与负极材料的复合在锂离子电池中的应用 ···· 120
  6.1.3 氧化石墨烯与正极材料的复合在锂离子电池中的应用 ···· 121
  6.1.4 氧化石墨烯作为导电添加剂在锂离子电池中应用 ···· 122
 6.2 氧化石墨烯在超级电容器中的应用 ·································· 123
  6.2.1 氧化石墨烯在双电层电容器中的应用 ···················· 123
  6.2.2 氧化石墨烯在赝电容器中的应用 ··························· 126
 6.3 氧化石墨烯在燃料电池中的应用 ······································ 128
 6.4 氧化石墨烯在太阳能电池中的应用 ·································· 129
 6.5 小结 ··· 130
 参考文献 ··· 131

第7章 氧化石墨烯在生物医学领域中的应用 /137

 7.1 氧化石墨烯的生物毒性及其机制 ······································ 137
  7.1.1 破坏细胞膜 ·· 139
  7.1.2 氧化应激 ·· 140

  7.1.3 其他因素 ·· 140
 7.2 氧化石墨烯在生物医学领域中的应用 ··· 140
  7.2.1 作为药物载体的应用 ·· 141
  7.2.2 抗菌 ·· 143
  7.2.3 生物成像 ·· 146
  7.2.4 生物传感器 ··· 147
  7.2.5 组织工程 ·· 148
 7.3 小结 ·· 149
 参考文献 ·· 149

第8章 氧化石墨烯在水处理中的应用研究 /155

 8.1 氧化石墨薄膜对溶液中离子的传输 ·· 155
 8.2 氧化石墨薄膜对气体和水分子的传输 ·· 159
 8.3 氧化石墨烯对重金属离子的吸附 ··· 161
  8.3.1 静电作用 ·· 161
  8.3.2 络合作用 ·· 161
  8.3.3 离子交换作用 ··· 162
  8.3.4 配体交换作用 ··· 162
 8.4 氧化石墨烯对水中有机污染物的吸附 ·· 163
 8.5 小结 ·· 165
 参考文献 ·· 166

第9章 氧化石墨烯在光电传感领域的应用 /170

 9.1 氧化石墨烯和还原氧化石墨烯的光学性质 ··· 171
  9.1.1 氧化石墨烯的光吸收特性 ··· 171
  9.1.2 氧化石墨烯的发光特性 ·· 172
  9.1.3 还原氧化石墨烯的光吸收特性 ······································ 173
  9.1.4 还原氧化石墨烯的发光特性 ··· 173
  9.1.5 氧化石墨烯量子点 ··· 173
  9.1.6 荧光淬灭 ·· 174
  9.1.7 光学非线性 ··· 174
 9.2 光电传感领域中的氧化石墨烯 ·· 175
  9.2.1 氧化石墨烯——光探测器件 ··· 175
  9.2.2 氧化石墨烯作为新型光源材料 ······································ 178
  9.2.3 氧化石墨烯-光纤传感器件 ··· 181
 9.3 小结 ·· 182
 参考文献 ·· 182

第1章
CHAPTER 1

氧化石墨烯的制备及表征

  2004 年，A. K. Geim 和 K. S. Novoselove 等在实验室对热解高定向石墨（Highly Oriented Pyrolytic Graphite，简称 HOPG）进行胶带微机械剥离，首次得到高质量石墨烯并观测到其室温量子霍尔效应。在这之后不久，石墨烯的研究热潮迅速主导了碳纳米材料领域，石墨烯成为了继碳纳米管（Carbon Nanotubes，简称 CNTs）和富勒烯（C_{60}）之后碳家族新的明星。随着石墨烯优异的电磁学性质和广泛的应用领域不断被揭示，多种制备石墨烯的方法也相继被研究和开发，例如：金属离子插层法[1]、液相剥离石墨法[2,3]、化学气相沉积（Chemical Vapor Deposition，简称 CVD）法[4]、真空碳化硅（SiC）外延生长法[5] 和化学还原氧化石墨烯（Graphene Oxide，简称 GO）法[6~8] 等。其中，氧化还原法的制备过程相对发展较早，生产效率较高，中间产品（如 GO）和不同还原程度的还原氧化石墨烯（Reduced Graphene Oxide，简称 RGO）易于调控且用途十分广泛，因此被认为是制备石墨烯最有效的途径之一。

  GO 被定义为具有单原子碳层的石墨氧化物，其前驱体材料氧化石墨合成历史由来已久。早在 1859 年，英国化学家 B. C. Brodie 用氯酸钾（$KClO_3$）和发烟硝酸（HNO_3）处理石墨，得到了分子式为 $C_{2.19}H_{0.80}O_{1.00}$ 的特殊物质，并将其命名为石墨烯酸或石墨氧化物[9]。直到进入 21 世纪，科学家认识到这种石墨氧化物是表面具有丰富含氧官能团和皱褶结构、厚度为单原子碳层的二维纳米片，更可以作为单原子层石墨——石墨烯的前驱体，因此将其命名为 GO。截至目前，已发现多种以石墨为原料制备 GO 的方法。通常先在硫酸、硝酸、磷酸或高氯酸等强酸中对石墨进行插层形成石墨层间化合物（Graphite Intercalation Compound，简称 GIC），再借助高锰酸钾、高铁酸钾等强氧化剂或电化学方法，将其氧化并剥离为 GO。在使用不同的氧化剂或氧化条件时，得到化合物的化学成分也略有不同。

本章中将主要介绍 GO 的合成方法、不同合成方法对其结构和化学组成的影响以及常用的表征方法。

1.1 合成

在本节中，我们将介绍不同的 GO 合成方法，并对它们的化学合成过程和产品形态进行分析和比较。这些方法包括：Brodie's 法，Staudenmaier's 法，Hummers' 法，Tour's 法，高铁酸钾氧化法，电化学法以及 GO 的其他制备方法。

1.1.1 Brodie's 法

1859 年，英国化学家 B. C. Brodie 在研究利用石墨的反应来探索石墨的结构时[9]，将质量比为 1∶3 的石墨与氯酸钾混合，并加入大量的发烟硝酸，60℃条件下反应 3～4 天，以水洗涤并干燥，然后真空干燥得到质量大于初始石墨原料质量的产物。经过分析，Brodie 认为产物由碳、氢、氧组成，重复氧化反应直至产物不发生任何变化得到极限氧化后 C∶H∶O 的组成为 61.04∶1.85∶37.11，可简化为分子式 $C_{2.19}H_{0.80}O_{1.00}$。这种产物可以被分散在中性或碱性的水中，但在酸性介质中絮凝，因此，Brodie 将这种物质称为"酸"。此外，他还发现 220℃下处理这种物质后 C∶H∶O 的比例变化为 80.13∶0.58∶19.29（$C_{5.51}H_{0.48}O_{1.00}$）。随着研究的深入，Brodie 确定这种物质的分子量为 33，并建议："因为它是一个独特的元素，这种形式的碳应该以一个名字来描述——Graphon。"

现在，石墨烯及二维材料的概念已经为大家所熟知。虽然，Brodie 当时得到的石墨分子式并不能真实地反映 GO 组分，但确实为制备具有单层碳原子厚度的 GO 提供了一种方法。之后，Botas 等[10]利用现代分析手段进行研究，发现 Brodie's 法处理石墨时对石墨的结构破坏较小，GO 产物中碳含量高达 70%～75%，其中 sp^2 的碳含量接近 40%；氧含量不足 30%，但主要以共轭的羟基和环氧基团存在；这些含氧官能团相对稳定，即使经过 2000℃高温处理也有 2.6% 的氧残留在样品中。当然，由于氯酸钾极易爆炸，实际研究中很少选择这种合成策略。

1.1.2 Staudenmaier's 法

为了简化反应步骤，Staudenmaier 对 Brodie's 法做了改进[11]。他在一个容器中分批次地将 $KClO_3$ 加入发烟 HNO_3 中，并且加入硫酸提高整个体系的酸浓度。最终产物的碳氧比为 2∶1，接近于 Brodie's 法得到的 2.19∶1。此外，Staudenmaier 的整个反应过程都是在一个容器中进行，操作起来比 Brodie's 法简便不少。然而，Staudenmaier's 法既耗时又危险：加入氯酸钾的步骤通常要持续 1 周以上，生成的二氧化氯需要用惰性气体进行特殊处理，而且在加料过程中随时可能发生爆炸。因为，Brodie's 法和 Staudenmaier's 法在合成的过程中都存在安全的不确定性，所以这两种合成策略并未得到广泛使用。

在 GO 的制备过程中，石墨原料是一个重要影响因素[12,13]。Chen 等[13]人报道了使用 5 个不同石墨供应商提供的石墨原料和使用改进的 Staudenmaier's 法制备 GO 产品的研究，发现虽然这些石墨样品具有相同的化学结构，但它们的晶粒尺寸、分散性、反应活性，特别是易氧化性有很大的差异。同时，由于这些晶体结构中活性位点的差异，导致最终 GO 产品的差异。此外，由于石墨原料的固有缺陷和结构的复杂性，这些反应的精确氧化机制很难进行阐明。

1.1.3 Hummers'法

1958 年，Hummers 和 Offeman 开发了一种替代 Staudenmaier's 法的 GO 制备方法。他们在浓硫酸、硝酸钠和高锰酸钾的无水混合物中，控制适当的反应温度进行石墨的氧化。反应的典型用量为：100g 石墨，50g 硝酸钠，300g 高锰酸钾和 2.3L 浓度为 98% 的浓硫酸。整个氧化过程在 2 小时内即可完成，所得产物的氧化水平与 Staudenmaier's 法相当[7]。通常认为，浓硫酸在无水条件下与 MnO_4^- 反应，生成氧化能力更强的 Mn_2O_7，因此反应过程中体系表现为墨绿色，参照反应式(1-1)和式(1-2)。Hummers 以硝酸钠替代了发烟硝酸，减少了反应过程中酸雾的生成，但制备过程中仍会产生诸如二氧化氮和氧化二氮的有毒气体[14]，然而硝酸钠在氧化过程的作用却无定论[15]。因此，近年来采用 Hummers'法制备 GO 时大都省去了硝酸钠[16]。

$$KMnO_4 + 3H_2SO_4 \longrightarrow K^+ + MnO_3^+ + H_3O^+ + 3HSO_4^- \quad (1-1)$$

$$MnO_3^+ + MnO_4^- \longrightarrow Mn_2O_7 \quad (1-2)$$

Fei 等[17]利用光学显微镜和拉曼光谱研究了 Hummers'法氧化石墨的机理，发现硫酸在数分钟内就可完全充满石墨层间形成石墨插层化合物，活性物质由外向内地氧化和剥离石墨片层，因此推测氧化反应主要受控于活性物质在石墨层间的扩散过程。早期以 Hummers'法制备的氧化石墨通常是核壳结构，由不完全氧化的石墨核和氧化石墨的壳层组成。为了提高收率并得到具有更高氧化程度的产品，Kovtyukhova 等[18]将石墨、浓硫酸、过硫酸钾和五氧化二磷研磨混合后在 80℃下进行数小时预氧化，经稀释、过滤、洗涤和干燥后，再进行 Hummers'法的氧化过程。如果石墨片层较小，或以预先经过热膨胀处理的膨胀石墨为原料，则可跳过预氧化处理过程。目前，这种改进的 Hummers'法成为研究者制备 GO 常用的手段之一。

由于 Hummers'法的氧化主要受氧化剂在石墨层间的扩散过程控制，剥离后的单原子层 GO 又容易被氧化切碎，因此很难获得大片径的、单层 GO。Dong 等[19]采用了先氧化、再分选、最后机械剥离的策略。他们先用双氧水和铬盐对天然石墨（100 目）进行预插层；用硫酸浸润，在冰浴条件下加入 2 倍高锰酸钾，35℃下反应 4 小时后，用 100 目筛网过滤反应产物；再将滤饼分散在 100mL 冰水和双氧水（以对过量的锰盐进行分解）中，利用自然沉降法（倒出上清液）洗涤氧化石墨 6~9 次；洗涤完成后以 800rpm 的磁力搅拌速度使氧化石墨在水中完全剥离。最终，他们得到了稳定的 GO 分散液，GO 片径约为 127μm。

Kim 等[20]人选取了三种不同的石墨原料，其横向尺寸分别为 0.75μm、1.22μm 和 1.65μm，长径比分别为 700、1200 和 1600。利用改进的 Hummers' 法，三种石墨原料的氧化产物的 C/O 比分别为 1.08、1.06、1.27；GO 产物在横向尺寸分布上有些许差异，由此导致了其向列相转变浓度在 0.53%~0.75%（质量分数）间波动。此外，Chen 等[15]证明了较小片径的石墨（3~20μm）比大的石墨（10~100μm）更容易被氧化，可使 110% 的产率提高到 171%。

1.1.4 Tour's 法

随着 GO 的研究不断推进，科学家们对 GO 合成方法的研究也越来越深入。2010 年，莱斯大学（Rice University）的 Tour 研究小组报道了一种新的 GO 合成方法，该方法避免了硝酸钠的使用，增加了高锰酸钾的用量，并引入了一种新的酸——磷酸[16]。其优势在于：合成的 GO 具有更完整的 sp^2 碳域[21]，如图 1.1 所示[21]，作者认为非晶磷环的形成有助于防止二醇被进一步氧化。

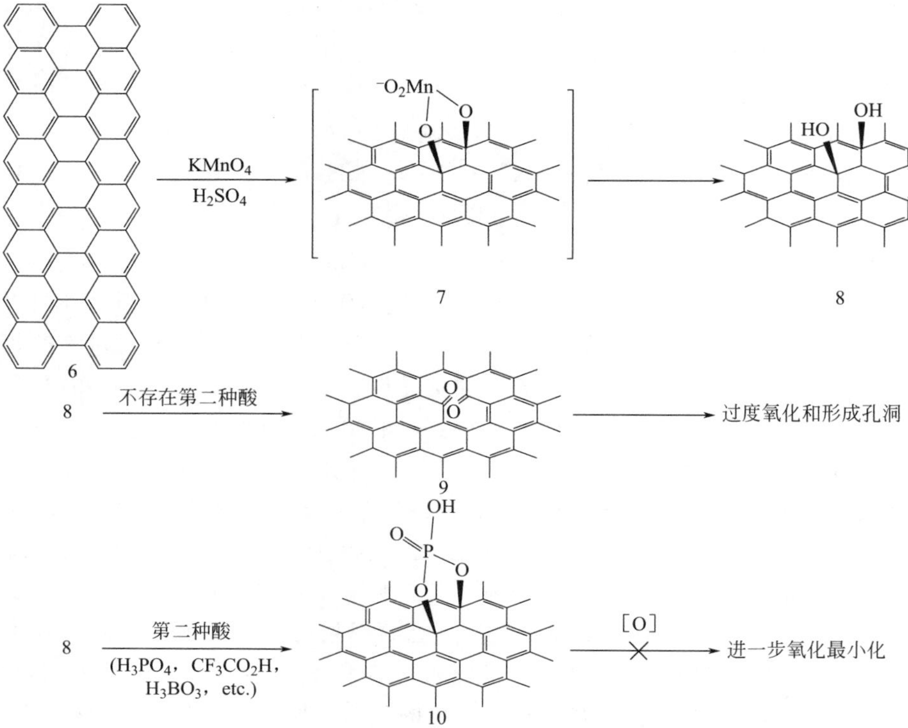

图 1.1 第二种酸对 sp^2 碳的过度氧化作用机理的推测[21]

在该方法中，石墨和高锰酸钾的比例为 1∶6，浓硫酸和磷酸的用量比例是 9∶1。因此通过这种方法制备的 GO 氧化程度比 Hummers' 法更高，甚至比改进的 Hummers' 法也要稍好一些。Tour 等认为，在反应体系中加入磷酸以后，磷酸基和石墨层基面上的邻位二醇结合形成五元环，这会使得更多的芳香族环产生，进而使反应获得的氧化石墨结构更加规整。这种合成方法对于石墨烯结构的破坏要明显小于 Hummers' 法。综合来说，这种方法的优势在于：不使用硝酸钠，不产生有毒气体（如二氧化氮、四氧化二氮、二氧化氯等）；反应过程中不涉及剧烈的放热，反应过程更加温和，更加环保；获得产物的氧化程度和产率都更高，结构更规则。显然，和 Hummers' 法相比，Tour's 法在大规模量产上更有优势，但含磷废酸的处理是该方法一个不容忽视的问题。

1.1.5　高铁酸钾氧化法

随着时间的推移，科学家们提出了很多对合成 GO 的各种优化和改进方法，尤其是很多对 Hummers' 法的优化。但是这些方法仍存在耗时长、制备过程复杂、反应残留的废物处理成本高等问题，大大地限制了氧化石墨的实际生产与应用。因此，仍迫切需要探究绿色（不含有毒气体和重金属污染）、安全（无爆炸性风险）、超快和低成本的合成方法。

高超研究组于 2015 年报道了一种新型、廉价、无毒的铁系氧化剂——高铁酸钾[22]，以取代沿用了半个多世纪的氯系、锰系氧化剂。首先，他们在室温条件下将 60g 高铁酸钾氧化剂加入 400mL 浓度为 93% 的硫酸中混匀。再将 10g 40 目石墨加入其中，并于室温下搅拌 1 小时。在硫酸的辅助下，高铁酸钾可以快速地插入石墨层间，使其被氧化并插层。在反应的过程中，高铁酸钾与浓硫酸反应产生氧气，从而加速石墨的插层，可将完全反应所需的时间缩短为 1 小时。利用高速离心将膏状沉淀用去离子水洗涤至接近中性，研究人员得到了品质较好的单层 GO。对比高铁酸钾和高锰酸钾分别作为氧化剂制备的氧化石墨样品，两者结构无明显差别。然而，高铁酸钾法避免了在制备过程和最终产品中引入污染重金属和有毒气体，而且硫酸可以回收，对环境友好。此外，通过这种方法制备的氧化石墨粉末在水中具有很高的溶解性，在水中分散后可以形成液晶，可以被加工成宏观的石墨烯纤维、薄膜和气凝胶。

为了提高氧化剂的利用效率，Yu 等[23] 人将高铁酸钾法与 Hummers' 法相结合，将石墨、氧化剂、硫酸的比例降至 1∶1.5∶10（质量∶质量∶体积）。他们的研究人员将 10g 磷片石墨、6g 高锰酸钾、4g 高铁酸钾和 100mL 浓硫酸混合，加入 0.01g 硼酸作为稳定剂，于低温（<5℃）下反应 1.5 小时。然后，再加入 5g 高锰酸钾，于 35℃ 下反应 3 小时。经过剥离和洗涤，得到了厚度为 2～3nm、横向尺寸为 10～20μm 的 GO 样品，并以其为原料制备了高性能的超级电容器电极材料。

Sofer 等[24]人对高铁酸钾氧化法的有效可行性提出了质疑。他们认为，高铁酸钾极易分解，市场上买到的高铁酸钾纯度与标签上严重不符（标称纯度90%，但实际最低只有20%），其中含有大量的氯酸钾。因此他们在实验室自制了纯度为76%的高铁酸钾，并与市售的高铁酸钾对比，分别在酸性（高铁酸钾极不稳定）、中性和碱性（高铁酸钾较稳定）条件处理石墨，制得产物的含氧量不足5%，且实验室自制的高铁酸钾的氧化产物仅含有2%的氧。经过分析，他们认为：高铁酸钾在与酸混合1小时内就基本完全分解，而市售高铁酸钾的氧化性则与其中的杂质（硝酸钾和氯酸钾）有关。

1.1.6 电化学制备法

传统生产 GO 的化学方法，如 Hummers'法，由于使用了具有危险性和爆炸性的化学物质，存在一系列的环境和安全问题。通过电化学氧化法制备 GO 可以避免这些问题，而且电化学氧化法具有反应简单、时间短的优点。

Ambrosi 等[25]人在双电极系统中以三种不同的物质（H_2SO_4、Na_2SO_4 及 $LiClO_4$）作为电解质，铂（Pt）箔作为负极，石墨纸作为正极，将石墨纸剥离来制备石墨烯，见图 1.2(a)。实验过程中，研究人员先施加+2V 的直流电压 2 分钟，然后施加+10V 直至石墨完全剥离或剥离不再发生。图 1.2(b) 和图 1.2(c) 分别是剥离前和施加+10V 长时间后石墨纸的照片，石墨纸的膨胀清晰可见。图 1.2(e)~(g) 中呈现的是初始阶段、施加电压（+10V）5 分钟后和 20 分钟后石墨纸的照片。剥离过程一旦完成（石墨纸完全消失或者没有进一步剥离达到稳定状态），通过过滤、洗涤收集溶液中剥离的石墨烯。图 1.2(d) 是该方法制备的石墨烯在 N,N-二甲基甲酰胺（N,N-Dimethyl Formamide，简称 DMF）中的稳定分散液，浓度为 1mg/mL。值得注意的是，使用 $LiClO_4$ 作为电解质，可以制备高含氧量的石墨烯材料，其 C/O 达到 4.0，接近某些 GO 中的碳氧比。因此，以 $LiClO_4$ 为电解质，通过电化学剥离法剥离石墨，可以获得 GO，是传统的化学氧化/剥离的替代方法，具有更加快速和安全的特点。

Cao 等[26]人报道了其利用电化学法制备 GO 的工作，提高了生产效率。该方法利用两步法制备 GO：a. 以石墨纸为正极、Pt 电极为负极、浓硫酸（>95%）为电解质，在 2.2V 电压、100mA 电流下进行 10 分钟的恒流充电，此时在石墨纸上形成第一阶段的 GIC；b. 以 GIC 为正极，Pt 电极为负极，0.1M 硫酸铵水溶液为电解质，在两端加载 10V 的电压，5~10 分钟后正极的 GIC 完全剥离，经过离心、洗涤、超声处理后得到 GO。这个过程中，硫酸电解质可以重复使用，直到完全被消耗，制备过程中没有其他污染物生成，是一种绿色、环保的制备方法；同时该方法制备氧化石墨烯具有高收率（产率>70%）、高质量（单层率达 90%）等特点，经过后处理后制备还原 GO，其电导率可达 54600S/m。

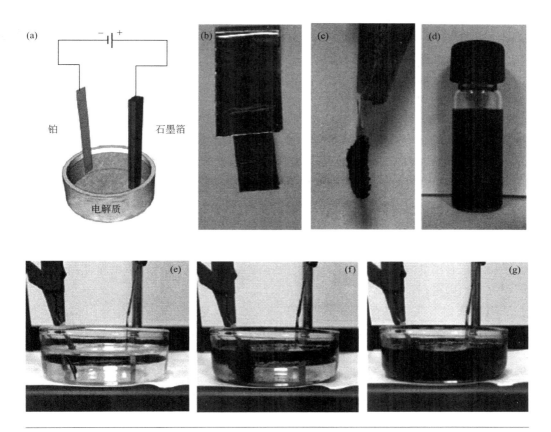

图 1.2 电化学方法制备实验过程[25]

Pei Songfeng 等[27]人报道了另一种电化学制备 GO 的方法。该方法以石墨的水电解氧化为基础。首先，以 Pt 为负极、柔性石墨纸为正极，在浓 H_2SO_4 中发生电化学插层，形成 I 型石墨插层化合物（Graphite Intercalation Compounds，简称 GICP）；然后，以 GICP 作为正极，在稀 H_2SO_4［50%（质量）］中进行电化学反应制备 GO。令人惊讶的是，蓝色的 GICP 在稀硫酸中几秒内即被快速氧化成黄色的氧化石墨；经抽滤和洗涤后，将滤饼的水溶液超声得到电化学合成的 GO（Electrochemical Synthesized Graphene Oxide，简称 EGO）分散液。

相比较其他的氧化方法，在电化学法制备氧化石墨的过程中，H_2SO_4 主要用以调节电化学反应以实现石墨烯的超快氧化；该方法不使用其他氧化剂，H_2SO_4 还可以回收，因此没有爆炸危险和金属离子污染。其次，电化学法的氧化速率比 Hummers' 法和 K_2FeO_4 方法氧化速率快 100 倍以上。另外，该电化学方法制备的氧化石墨清洗更容易，清洗水用量比普通 Hummers' 法大为减少，这是因为在 Hummers' 法中，氧化石墨与浓硫酸和其他氧化剂均匀混合，形成非常黏稠的浆料，洗涤需要更多的水。该方法还具有良好的可控性，通过改变电化学氧化过程中 H_2SO_4 溶液的浓度，可以容易地调节氧化石墨片的氧化度，并且可以通过超声时间调节层数和尺寸。通过将 GICP 连续引入稀 H_2SO_4 中，有可能实现氧化石墨片材的连续生产。总结而言，该电化学氧化方法具有较高的安全性、超快速合成、易于控制、环保、无金属离子污染和易于扩大的优点。

1.1.7 其他氧化石墨烯的制备方法

阳离子纯化是另一种制造氧化石墨的方法之一，但是其制备过程烦琐，每一步都需要进行长时间的清洗、过滤、离心和透析。此外，被钾盐污染的土壤具有很高的流动性，对人体健康构成一定危害。而且，在水冲洗过程中由于氧化石墨体积的膨胀和氧化石墨的凝胶化，极大地减慢洗涤的效率；Kim 等人[28] 提出使用盐酸和丙酮取代水来进行洗涤。

随着氧化石墨以及石墨烯研究的发展，Mhamane 等人[29] 选取天然植物为原料，通过多次洗涤，在 60℃下烘干 3 天，把植物磨成粉末后充当石墨原料，再通过改进 Hummers' 法来制备 GO。该方法采用高锰酸钾和硫酸相结合的方法。虽然高锰酸钾是一种常用的氧化剂，但其活性物质实际上是一种抗氧化剂，在该反应中利用高锰酸钾与硫酸反应形成的具有强化性的氧化物来氧化原料，该氧化物在加热到高于 55℃温度时或者与有机化合物接触时就会发生爆炸[30,31]。

1.2 氧化石墨烯的结构研究

为更好地应用 GO，了解 GO 的结构是必要的。GO 主要是由碳、氧和氢元素组成，碳氧比保持在 1.5～2.5，在过去的 150 年中，GO 确切化学结构的探究一直是相关研究者致力的方向。已经被报道的结构有：Hofmann、Ruess、Scholz-Boehm、Nakajima-Matsuo、Lerf-Klinowski、Dékány 和 Ajayan 模型，参照图 1.3。

第一个氧化石墨模型是由 Hofmann 提出的[32]。他们认为 GO 的结构应该是由 1,2-环氧化合物这一单元重复出现在 GO 的基准面上所组成。1946 年，Ruess 提出了一种由 sp^3 为杂化基准面的新模型，而不是 Hofmann 和 Holst 提出的以 sp^2 杂化为基准面的模型[33]。该模型由 1,3-环氧化物和羟基为基准单元共同组成，这个模型解释了为什么在氧化石墨中有氢原子的存在。20 年后（1969 年），Scholz 和

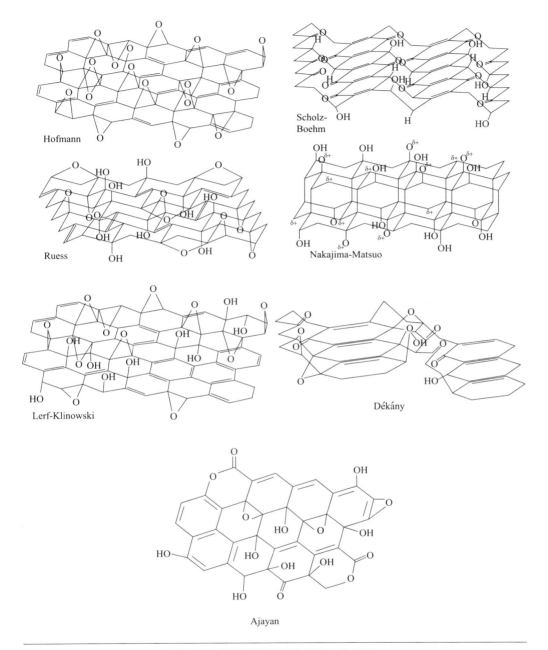

图 1.3 氧化石墨的几种假设结构模型

Boehm 提出了一个由羟基和酮共同组成的新模型。随后由 Nakajima 和 Matsuo 提出了一种与插层石墨非常相似的结构模型[34]。早期的这些 GO 模型的构建主要是基于元素分析、化学反应和X射线衍射（X-ray Diffraction，简称XRD）等研究结果。

随后，Lerf和Klinowski通过同位素标记，结合魔角旋转核磁共振（Magic Angle Spinning-Nuclear Magnetic Resonance，简称MAS-NMR），在两个氧化石墨上进行^{13}C和^{1}H标记，研究得到了"Lerf和Klinowski模型"[35,36]。该模型中氧化石墨被假定为是含有未被氧化苯环的芳香区和另一个有脂肪六元环的区域所组成。含氧的官能团，如1,2-环氧化物和羟基填充在基准面上，羧基和羟基主要集中在基准面的边缘处。目前，使用高分辨透射电子显微镜（High Resolution Transmission Electron Microscope，简称HRTEM）观察氧化石墨表面得到的数据证明氧化石墨薄片上确实存在Lerf-Klinowski模型所表述的特征[37]。

基于元素分析、X射线光电子能谱、漫反射红外傅里叶变换光谱学、电子自旋共振、透射电子显微镜（Transmission Electron Microscope，简称TEM）、XRD和^{13}C MAS NMR的研究[38]，在Ruess和Scholz-Boehm模型的基础上，Dékány提出了氧化石墨是由椅子型反式链环己烷和波纹六角带所组成。环己烷是由1,3-环氧化物和叔羟基组成，六边形带由环酮和醌类组成。此外，模型中还引入了酚类基团来解释氧化石墨的酸性来源。

在一次确定氧化石墨的详细结构的尝试中，Ruoff和他的同事合成了一种由100%^{13}C标记的氧化石墨，用^{13}CSS NMR（Solid-state nuclear magnetic resonance）来进行分析[39]。测试表明，大多数的羟基和环氧化合物都是与碳相连。此外，此研究还证明了羰基是不与sp^2的碳、羟基和环氧相连接的。在100ppm浓度下，可以检测到非质子化碳的信号存在，但并没有特别指出其具有何种作用。尽管如此，从核磁共振（Nuclear Magnetic Resonance，简称NMR）分析中得到的数据表明，氧化石墨可能的结构模型只有两种：Lerf-Klinowski和Dékány模型。

Ajayan和他的同事在随后的工作中表明，在^{13}C MAS NMR中100ppm处的信号来自于乳糖，特别是2-羟基苯酐或1,3-二羟基黄酮[40]。氧化石墨上官能团的相对比例为：115（羟基和环氧化合物）:3（乳糖O-C-O）:63（石墨化sp^2碳）:10（乳酸+酯酸+羧基）:9（酮羰基）。

在确定氧化石墨的结构细节方面，Tour和同事采取了不同的方法。在石墨氧化终止阶段使用非水溶剂替代水，生产出由环氧基团主导的"GO"[41]。在通常使用水终止氧化反应的合成中，少量的共价硫酸盐和羟基与水发生大量的化学反应，从而影响氧化石墨的酸性。事实上，氧化石墨的酸性是边缘处于平衡状态的水化物和不完全水解的硫酸盐所形成的酮共同引起的。这个解释与"酸性来源于羧基"的经典解释形成了对比。此外，位于100ppm位置处的^{13}C MAS NMR的信号被归因于半缩醛。随后，该小组对氧化石墨的酸性特性进行更深入地研究，发现在氧化石墨上的酸性官能团（即乙烯基酸）是通过与水的相互作用而逐渐产生的。有人提出,氧化石墨的结构不是一种具有固定官能团的"静态结构",而是在水的存在下不断发生变化的"动态结构模型"。

综上所述,几种不同制备 GO 的方法其目标都是使 GO 的制备过程能够更简单、效率更高、产量更大、过程更安全,产物质量更加可控。同时,石墨原料对氧化石墨合成过程具有一定影响,氧化石墨结构也被逐渐深入认知。虽然现在 GO 制备已经实现规模化,但仍然缺乏对 GO 氧化过程和机理的详细认识,这也阻碍了相关化学工程的发展以及对各种 GO 控制与应用的探索。因此对于一些 GO 合成相关的关键技术问题,例如,尺寸分布控制、边缘结构选择、官能团种类和位置控制等应继续进行深入研究。

1.3 氧化石墨烯的光谱特征和结构表征方法

GO 是一种独特的氧化碳结构,它超出了有机化合物和多环芳烃的范围,因此描述这种特殊结构非常有趣和具有挑战性。在这一节中,我们将从光谱分析方法来阐明它的分子结构。

1.3.1 氧化石墨烯的光谱特征

(1) 固体核磁共振 (Solid State Nuclear Magnetic Resonance,简称 SSNMR) 光谱分析

SSNMR 光谱分析认为 GO 薄片是属于类似于胶体的巨大分子。因为 GO 在水中的分散度低,较难用液相 NMR 来分析。在文献中,表征 GO 最强大、最精确的技术是通过同位素 ^{13}C 来进行 GO SSNMR 光谱的测定[36]。由于 ^{13}C(1.1%)在自然条件中含量较低,在样品的测量过程中所产生的信噪比低,需要通过长时间的测试以得到高质量的数据。因此,2008 年 Ruoff 和他的同事通过使用同位素 ^{13}C 对 GO 进行标记,将 GO 的结构研究推上了一个新的高度[39]。根据研究分析,在有 ^{13}C 标记的氧化石墨 NMR 谱中,交叉极化/魔角旋转(Cross Polarization Magic Angle Spinning,简称 CP/MAS)显示出 60ppm、70ppm 和 130ppm 的三个宽共振。此外,虽然利用同位素标记法可以更好地通过 SSNMR 光谱对氧化石墨进行分析,但在 SSNMR 光谱图中仍然还存在一些未知的峰。随着研究的深入,科学家们对这些未知峰进行了相关的研究并给出了详细的解释。例如:位于 SSNMR 光谱 101ppm 的峰就是一个长期以来被忽略的峰位置。相关研究认为,101ppm 的峰可能来自图 1.4 所示的五元环或六元内酰环的乳糖醇结构[40]。

另外,还可以利用 SSNMR 光谱表征同位素 ^{13}C 在二维材料中是双量子或单量子(2Q/SQ),亦可通过在 SSNMR 光谱中观察同位素 ^{13}C 在二维材料中化学位移的各向异性以及同位素 ^{13}C 在二维材料中是三量子还是单量子(3Q/SQ)[42]。

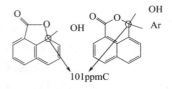

图 1.4　五元和六元内酰环的结构
（用红色圈出的碳是在 101ppm 处产生 ^{13}C NMR 信号的碳）

2Q/SQ 光谱消除了二维光谱中的对角信号,很清晰地呈现出了 ^{13}C-OH 中的碳和 ^{13}C-O-^{13}C 中的碳信号。3Q/SQ 光谱提供了三个不同碳原子之间的相互关系,并且提供了大量关于不同功能化碳原子以及氧化石墨表面官能团的分布情况的相关信息。

值得注意的是,大部分关于 SSNMR 光谱所用到的 GO 都是通过改进 Hummers' 法制备得到的,而其他制备方法得到的 GO 产品在官能团的相对含量上会有不同程度的差异[16]。

（2）漫反射傅里叶红外光谱（Diffuse Reflectance Infrared Fourier Transform Spectroscopy,简称 DRIFT）表征

除了 SSNMR 外,DRIFT 是另一种检测氧化碳化学结构的有力工具[38,43]。GO 样品具有很强的吸湿性——有高达 25% 质量比的水分子被吸附在 GO 的结构中,而高的含水量是 DRIFT 分析过程中的最大障碍。因此,Decany 和同事报道了一种采用 Brodie's 法制备的 GO,并用同位素 D（氘）标记后进行 DRIFT 分析的方法,成功地对 GO 的各个 DRIFT 峰进行了区分[38,43]。

（3）拉曼光谱、紫外-可见分光光度计、XRD 和 X 射线光电子能谱（X-Ray Photoelectron Spectroscopy,简称 XPS）表征

作为一个部分有序结构,GO 表面含有各种不同的含氧官能团结构,借助拉曼光谱、扫描电子显微镜[44,45]和透射电子显微镜[46,47],我们可以很容易地观察到有部分 sp^3 的碳被分离出来形成 sp^2 的碳,并且对这一过程进行了研究[48,49]。

在拉曼光谱中（如图 1.5 所示）,GO 的特征峰主要是由广义 D 峰（约 $1360cm^{-1}$）和 G 峰（约 $1580cm^{-1}$）以及位于 $2681\sim3050cm^{-1}$ 所形成的宽峰组成。D 峰/G 峰比约为 0.95,说明在 GO 的晶格中存在着大量的缺陷。然而,在对 GO 进行化学还原后,通常会观察到 D 峰/G 峰比值的增加,此现象和还原过程中脱氧引起的拓扑缺陷增加有关,但细节仍然需要进一步的研究。

在紫外-可见分光光度计的光谱测定中,GO 在水中含有两个特征峰,一个是位于 233nm 波长处由 C=C 所形成的 π-π^* 跃迁所产生的吸收峰,另一个是位于

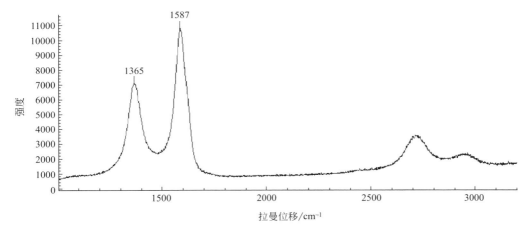

峰号	中心	高	宽	面积	绝对强度	低边缘	高边缘
1	1364.62	5206.39	64.8991	1.30113e+006	7194.66	1200.56	1455.64
2	1586.61	9247.06	64.3208	1.45322e+006	11009.9	1473.96	1708.19

图 1.5 GO 的一个典型拉曼光谱图

290～300nm 处由 C=O 所形成的 n-π^* 跃迁所产生的吸收峰[50]。在对 GO 进行还原后，我们通常会观察到位于 233nm 处的吸收峰发生红移，位于 290～300nm 处的峰消失。

利用 XRD 来分析 GO 一个位于 11°的宽峰，发现其受 GO 的氧化程度、水化程度和所处环境的湿度等共同影响。已知报道的 GO 样品层间距离从 5.97Å[43] 到 9.5Å[16] 不等。

在 XPS 图谱中 GO 会以 284eV（sp^2 碳）和 286eV（氧化碳）处为中心形成 2 个加宽峰或重叠的 GO 碳 1s1s 特征峰。普通情况下，XPS 无法严格区分诸多碳含氧官能团的精确位置，这使得通过利用 XPS 来准确分析 GO 和 RGO 官能团种类和数量存在着很大的不确定性。

1.3.2 氧化石墨烯的形态学特征及热稳定性

研究人员通常采用 TEM 和扫描透射电子显微镜（Scanning Transmission Electron Microscope，简称 STEM）来表征 GO 基面和边缘局部环境等微观结构。原子力显微镜（Atomic Force Microscope，简称 AFM）在许多文献中也被用来证明 GO 薄片的层数（单层或少层）和厚度。扫描隧道显微镜（Scanning Tunneling Microscope，简称 STM）也被用来表征 GO，但是由于 GO 缺乏导电性，导致 STM 在某些情况下无法对 GO 进行分析。

(1) TEM 表征

随着石墨烯结构研究的深入，GO 的二维结构也受到了人们的广泛关注[48,51]。HRTEM 或 STEM 成像技术尽管在文献报道之间还存在差异，但仍为我们更好地了解 GO 的原子结构和电子束辐照线的稳定性提供了参考依据。例如，有的文献报道是在 80kV 电子束下收集到稳定的 GO 图像[45]而在某些情况下只能使用最多 60kV 的电子束来对 GO 进行成像[52]。在这里我们展示通过 Hummers' 法制备得到的氧化石墨在不同放大倍数下的 TEM 图像，如图 1.6 所示。大量 GO 的 TEM 图像也已被报道[53]。人们通过 TEM 对 GO 进行表征，观察到了 sp^2 碳区域和附着的含氧基团。

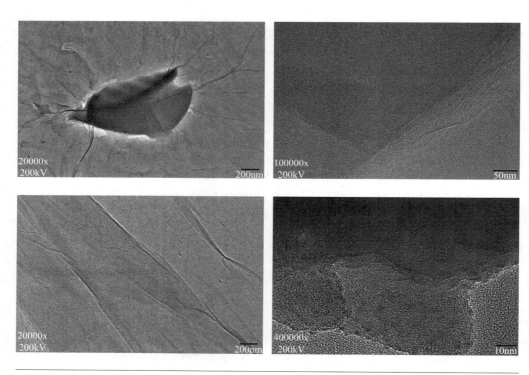

图 1.6 用改良的 Hummers' 方法制备的氧化石墨样品在不同放大倍率下的 TEM 图像

(2) AFM 和 STM 表征

AFM 通常被用来测量 GO 的厚度和表征 GO 薄片的横向形貌。研究者们通常使用溶剂对 Si 表面进行预处理，使其具有超亲水性，以便于 GO 悬浮液在其表面形成涂层。图 1.7 显示了云母表面涂覆 GO 后形成的典型 GO 的 AFM 图像。可以清晰地看出，GO 的厚度约为 1.2nm。由于 GO 的上下表面具有大量的含氧官能团，导致 GO 的厚度要比石墨中原子间距 3.4Å 大三倍。虽然 AFM 测试的最高分辨率高达 1Å[54]，但非常精确或可靠的测量数据并不容易得到。很少有报道还原后的 GO 厚度有减少的现象[55]。

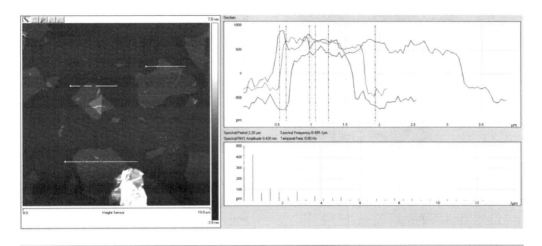

图 1.7　GO 片在云母基底上 AFM 图像（a）以及对应
(a) 图中标记区域 AFM 图像的对应厚度（b）

另一方面，由于 GO 的绝缘特性，用 STM 来表征 GO 的报道较少。然而 RGO 具有更强的导电性，使得我们可以用 STM 来表征 RGO 的图像。如图 1.8 所示，从 STM 图像中可以发现，GO 的表面形貌具有非周期性的随机分布特征，很难将其与之前得到的高分辨率 TEM 图像相关联。

（3）热重分析（Thermogravimetric Analysis，简称 TGA）和差示扫描量热法（Differential Scanning Calorimetry，简称 DSC）表征

GO 在受热的条件下不稳定，TGA 一般被用来表征 GO 的热稳定性。当 GO 在氩气或氮气的条件下加热时，在 60~80℃ 之间缓慢分解。当加热温度达到 950℃ 时，GO 会失去自身 60% 的重量。如图 1.9 所示，在此过程中也会发生吸附水的流失。值得注意的是，在对 GO 进行 TGA 分析时，升温速率一般设置为 1℃/min，以防止气体迅速放出，导致爆炸造成危险。

如图 1.10 所示，还可以把 TGA 和 DSC 结合起来对 GO 进行分析。在氮气的条件下对 GO 进行加热，加热速率为 10℃/min 时，可以看到在 DSC 中出现了第一个水分解产生的峰，GO 的失重约为 17%。当温度达到 160~200℃ 时，在 TGA 和 DSC 图谱都显现出 GO 失去了大量的重量。相应地，在 DSC 图谱中位于 180℃ 处显示出了一个高度集中的峰，这是由于不稳定的含氧官能团的分解所产生。

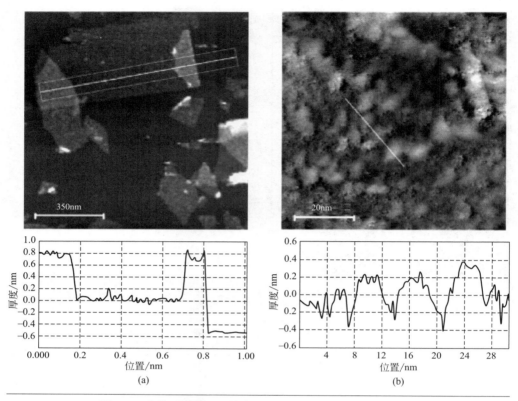

图 1.8 化学还原的 GO 纳米片沉积到云母片上的 STM 图像

（a）石墨烯的片径大小和 GO 的厚度；(b)（a）图标记区域的放大图像[45]

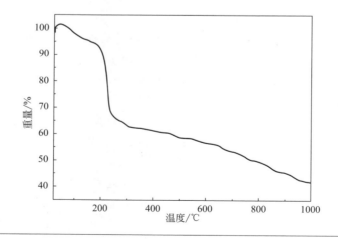

图 1.9 氧化石墨的 TGA 测试

（升温速率：2℃/min 在超高纯度 N_2 氛围中。样品在加热至较高温度之前在 25℃ 的 N_2 气氛保持 300min 以除去所有物理吸收的水）

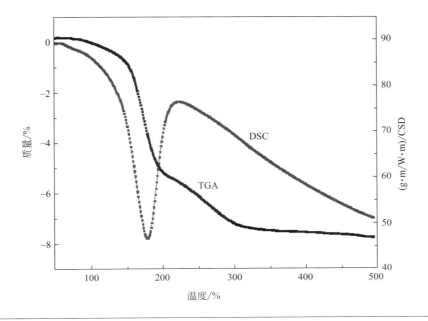

图 1.10　氧化石墨的热重-差示量热扫描曲线

1.4　小结

GO 作为一个半世纪前发现的巨大有机分子,虽然在过去十年中主要作为石墨烯的重要前驱体出现,但现在它已经独立于石墨烯之外有了自己的历史。

在这一章中,我们总结并讨论了 GO 的合成、表征、结构,并试图以化学和材料科学为基础来理解 GO。长期研究表明,从天然石墨材料出发,经强化学氧化可大规模制备 GO。通过各种特征分析证明 GO 是一种具有二维结构的氧化碳化合物,或是一种上下两面都含有大量含氧基团所形成的六边形环状结构。在 GO 中碳原子以 sp^2 和 sp^3 的结构混合在一起,使得 GO 同时具有波纹状、两亲性和荧光性以及含氧官能团带来的吸湿性、分散性和化学反应性。GO 的单层原子二维特性决定了 GO 薄膜透光性良好,并可以进一步地被还原成透明的、导电的薄膜。目前,GO 已经被广泛应用于多个不同的领域,成为基础科学和现代技术的重要材料。

参考文献

[1] Viculis Lisa M, Mack Julia J, Kaner Richard B. A chemical route to carbon nanoscrolls [J]. Science, **2003**, 299 (5611): 1361.

[2] Coleman Jonathan, Lotya Mustafa, Gallagher Arlene, et al. Two-Dimensional Nanosheets Produced by Liquid Exfoliation of Layered Materials [J]. Cheminform **2011**, 42 (18): 568-571.

[3] Yenny Hernandez, Valeria Nicolosi, Mustafa Lotya, et al. High-yield production of graphene by liquid-phase exfoliation of graphite [J]. Nature Nanotechnology, **2008**, 3 (9): 563-568.

[4] Li Xuesong, Cai Weiwei, An Jinho, et al. Large-area synthesis of high-quality and uniform graphene films on copper foils [J]. Science, **2009**, 324 (5932): 1312-1314.

[5] Berger Claire, Song Zhimin, Li Xuebin, et al. Electronic confinement and coherence in patterned epitaxial graphene [J]. Science, **2006**, 312 (5777): 1191-1196.

[6] Gilje Scott, Han Song, Wang Minsheng, et al. A chemical route to graphene for device applications [J]. Nano Letters, **2007**, 7 (11): 3394-3398.

[7] Stankovich Sasha, Dikin Dmitriy A., Piner Richard D., et al. Synthesis of graphene-based nanosheets via chemical reduction of exfoliated graphite oxide [J]. Carbon, **2007**, 45 (7): 1558-1565.

[8] Petr Šimek, Kateřina Klimová, David Sedmidubský, etc.. Towards graphene iodide: iodination of graphite oxide [J]. Nanoscale, **2015**, (7), 261-270.

[9] Brodie Benjamin Collins. On the Atomic Weight of Graphite [J]. Philosophical Transactions of the Royal Society of London, **2009**, 149 (1): 249-259.

[10] Botas Cristina, Alvarez Patricia, Blanco Patricia, et al. Graphene materials with different structures prepared from the same graphite by the Hummers and Brodie methods [J]. Carbon, **2013**, 65 (6): 156-164.

[11] Staudenmaier L. Verfahren zur Darstellung der Graphitsäure [J]. European Journal of Inorganic Chemistry, **1898**, 31 (2): 1481-1487.

[12] Botas Cristina, Álvarez Patricia, Blanco Clara, et al. The effect of the parent graphite on the structure of graphene oxide [J]. Carbon, **2012**, 50 (1): 275-282.

[13] Chen Zhi Li, Kam Fong Yu, Goh G S., et al. Influence of Graphite Source on Chemical Oxidative Reactivity [J]. Chemistry of Materials, **2013**, 25 (15): 2944-2949.

[14] Zaaba N. I., Foo K. L., Hashim U., et al. Synthesis of Graphene Oxide using Modified Hummers Method: Solvent Influence [J]. Procedia Engineering **2017**, 184: 469-477.

[15] Chen Ji, Li Yingru, Huang Liang, et al. High-yield preparation of graphene oxide from small graphite flakes via an improved Hummers method with a simple purification process [J]. Carbon, **2015**, 81 (1): 826-834.

[16] Marcano Daniela C, Kosynkin Dmitry V, Berlin Jacob M, et al. Improved synthesis of graphene oxide [J]. Acs Nano, **2010**, 4 (8): 4806.

[17] Fei Huilong, Ye Ruquan, Ye Gonglan, et al. Boron- and nitrogen-doped graphene quantum dots/graphene hybrid nanoplatelets as efficient electrocatalysts for oxygen reduction [J]. Acs Nano, **2014**, 8 (10): 10837.

[18] Kovtyukhova Nina I., Ollivier Patricia J., Martin Benjamin R., et al. Layer-by-Layer Assembly of Ultrathin Composite Films from Micron-Sized Graphite Oxide Sheets and Polycations [J]. Chemistry of Materials, **1999**, 11 (3): 771-778.

[19] Dong Lei, Chen Zhongxin, Lin Shan, et al. Reactivity-Controlled Preparation of Ultralarge Graphene Oxide by Chemical Expansion of Graphite [J]. Chemistry of Materials, 2017, 29 (2): 564-572.

[20] Ji Eun Kim, Han Tae Hee, Lee Sun Hwa, et al. Graphene Oxide Liquid Crystals † [J]. Angewandte Chemie International Edition, 2011: 3043-3047.

[21] Higginbotham Amanda L, Kosynkin Dmitry V, Alexander Sinitskii, et al. Lower-defect graphene oxide nanoribbons from multiwalled carbon nanotubes [J]. Acs Nano, 2010, 4 (4): 2059-2069.

[22] li Peng, xu Zhen, Liu Zheng, et al. An iron-based green approach to 1-h production of single-layer graphene oxide [J]. Nature Communications, 2015, 6: 5716.

[23] Yu Huitao, Zhang Bangwen, Bulin Chaoke, et al. High-efficient Synthesis of Graphene Oxide Based on Improved Hummers Method [J]. Scientific Reports, 2016, 6: 36143.

[24] Sofer Zdeněk, Luxa Jan, Jankovský Ondřej, et al. Synthesis of Graphene Oxide by Oxidation of Graphite with Ferrate (VI) Compounds: Myth or Reality? [J]. Angewandte Chemie 2016, 55 (39): 11965-11969.

[25] Ambrosi Adriano, Pumera Martin. Electrochemically Exfoliated Graphene and Graphene Oxide for Energy Storage and Electrochemistry Applications [J]. Chemistry of Materials, 2016, 22 (1): 153-159.

[26] Cao Jianyun, He Pei, Mohammed Mahdi A., et al. Two-step electrochemical intercalation and oxidation of graphite for the mass production of graphene oxide [J]. Journal of the American Chemical Society, 2017, 139 (48): 17446-17456.

[27] Pei Songfeng, Wei Qinwei, Huang Kun, et al. Green synthesis of graphene oxide by seconds timescale water electrolytic oxidation [J]. Nature Communications, 2018, 9 (1): 145.

[28] Kim Franklin, Luo Jiayan, Cruz-Silva Rodolfo, et al. Self-Propagating Domino-like Reactions in Oxidized Graphite [J]. Advanced Functional Materials, 2010, 20 (17): 2867-2873.

[29] Mhamane Dattakumar, Ramadan Wegdan, Fawzy Manal, et al. From graphite oxide to highly water dispersible functionalized graphene by single step plant extract-induced deoxygenation [J]. Green Chemistry, 2011, 13 (8): 1990-1996.

[30] Koch Klaus R.. Oxidation by Mn_2O_7: An impressive demonstration of the powerful oxidizing property of dimanganeseheptoxide [J]. Journal of Chemical Education, 1982, 59 (11): 973.

[31] Bargar Keith E., Erd Richard C., Keith Terry E. C., et al. Dachiardite from Yellowstone National Park Wyoming [J]. Canadion Mineralogist, 1987, 25: 475-480.

[32] Hofmann U. Graphit und Graphitverbindungen [M]. Berlin Heidelberg, Springer, 1939.

[33] Ruess G.. Über das Graphitoxyhydroxyd (Graphitoxyd) [J]. Monatshefte Für Chemie Und Verwandte Teile Anderer Wissenschaften, 1947, 76 (3-5): 381-417.

[34] Nakajima Tsuyoshi, Matsuo Yoshiaki. Formation process and structure of graphite oxide [J]. Carbon, 1994, 32 (3): 469-475.

[35] He Heyong, Riedl Thomas, Lerf Anton, et al. Solid-State NMR Studies of the Structure of Graphite Oxide [J]. Journal of Physical Chemistry B, 1996, 100 (51): 19954-19958.

[36] Lerf Anton, He Heyong, Forster Michael, et al. Structure of Graphite Oxide Revisited [J] Journal of Physical Chemistry B, 1998, 102 (23): 4477-4482.

[37] Kris Erickson, Rolf Erni, Zonghoon Lee, et al. Determination of the local chemical structure of graphene oxide and reduced graphene oxide [J]. Advanced Materials, 2010, 22 (40): 4467-4472.

[38] Szabó Tamás, Berkesi Ottó, Forgó Péter, et al. Evolution of Surface Functional Groups in a Series of Progressively Oxidized Graphite Oxides [J]. Chemistry of Materials, 2015, 18 (11): 2740-2749.

[39] Cai Weiwei, Piner Richard, Stadermann Frank, et al. Synthesis and solid-state NMR structural charac-

terization of 13C-labeled graphite oxide [J]. Science, **2008**, 321 (5897): 1815-1817.

[40] Gao Wei, Alemany Lawrence B., Ci Lijie, et al. New insights into the structure and reduction of graphite oxide [J]. Nature Chemistry, **2009**, 1 (5): 403-408.

[41] Dimiev Ayrat, Kosynkin Dmitry V., Alemany Lawrence B., et al. Pristine Graphite Oxide [J]. Journal of the American Chemical Society, **2012**, 134 (5): 2815-2822.

[42] Casabianca Leah, Shaibat Medhat, Cai Weiwei, et al. NMR-Based Structural Modeling of Graphite Oxide Using Multidimensional C-13 Solid-State NMR and ab Initio Chemical Shift Calculations [J]. Journal of the American Chemical Society, **2010**, 132: 5672-5676.

[43] Szabó Tamás, Berkesi Ottó, Dékány Imre. DRIFT study of deuterium-exchanged graphite oxide [J]. Carbon, **2005**, 43 (15): 3186-3189.

[44] Ishigami Masa, Chen J. H., Cullen William, et al. Atomic structure of graphene on SiO_2 [J]. Nano Letters, **2007**, 7 (6): 1643-1648.

[45] Paredes J I, Villar-Rodil S, Solís-Fernández Pablo, et al. Atomic force and scanning tunneling microscopy imaging of graphene nanosheets derived from graphite oxide [J]. Langmuir the Acs Journal of Surfaces Colloids, **2009**, 25 (10): 5957-5968.

[46] Navarro Cristina, C Meyer Jannik, Sundaram Ravi, et al. Atomic structure of reduced graphene oxide [J]. Nano Letters, **2010**, 10 (4): 1144-1148.

[47] Wilson Neil, Pandey Priyanka, Beanland Richard, et al. Graphene oxide: structural analysis and application as a highly transparent support for electron microscopy [J]. Acs Nano, **2009**, 3 (9): 2547-2556.

[48] Eda Goki, Mattevi Cecilia, Yamaguchi Hisato, et al. Insulator to semi-metal transition in graphene oxide [J]. Physics, **2009**, 113 (35): 15768-15771.

[49] Kaiser Alan, Navarro Cristina, Sundaram Ravi, et al. Electrical conduction mechanism in chemically derived graphene monolayers [J]. Nano Letters, **2009**, 9 (5): 1787-1792.

[50] Luo Zhengtang, Lu Ye, Somers Luke A, et al. High yield preparation of macroscopic graphene oxide membranes [J]. Journal of the American Chemical Society, **2009**, 131 (3): 898-899.

[51] Mkhoyan K, Andre Mkhoyan K, Contryman Alexander W., et al. Atomic and electronic structure of graphene-oxide [J]. Microscopy Microanalysis, **2009**, 16 (S2): 1704-1705.

[52] Gao Wei, Wu Gang, Janicke Michael, et al. Ozonated graphene oxide film as a proton-exchange membrane [J]. Angew Chem Int Ed Engl, **2014**, 53 (14): 3588-3593.

[53] Pacilé D, Meyer J, Rodriguez Arantxa, et al. Electronic properties and atomic structure of graphene oxide membranes [J]. Carbon **2011**, 49 (3): 966-972.

[54] Alexander S., Hellemans L., Marti Othmar, et al. An atomic-resolution atomic-force microscope implemented using an optical lever [J]. Journal of Applied Physics, **1989**, 65 (1): 164-167.

[55] Ma Chen, Chen Zhongxin, Fang Ming, et al. Controlled synthesis of graphene sheets with tunable sizes by hydrothermal cutting [J]. Journal of Nanoparticle Research, **2012**, 14 (8): 996.

第 2 章
CHAPTER 2

氧化石墨的还原

　　石墨烯从结构上而言是一种由碳原子以 sp^2 杂化轨道组成的六角蜂巢型二维晶格。由于碳碳共轭结构的存在，石墨烯在大面积上保持均匀的导电、强度、原子级平整及其他特性。第 1 章介绍了不同氧化技术制备氧化石墨的方法，但是含氧官能团的引入破坏了石墨烯的共轭结构，因此氧化石墨呈现电绝缘或者半导体特性，颜色为棕色或者黄色。为恢复共轭特性（如导电性），需要去除或者部分去除氧化石墨中的氧，该过程称为氧化石墨的还原。氧化石墨还原过程和氧化石墨的剥离过程经常同时发生，但是也可先剥离成 GO 再进行还原。以氧化石墨为原料，研究者们发展了热还原、电化学还原、化学还原等方法以去除氧化石墨上的含氧基团，得到具有不同性质（电子结构、物理和表面形态）的石墨烯（有时被称为 RGO）。通过这些方法制备的石墨烯材料在结构上通常存在拓扑缺陷或保留有杂质原子，但石墨烯的基本结构和若干特性仍得以恢复，因此通过还原氧化石墨制备石墨烯是规模化生产石墨烯和大范围应用石墨烯的重要途径。下面将对氧化石墨的还原方法及还原机理进行总结。

2.1　氧化石墨烯还原评价方法

　　由于氧化石墨在还原过程中微观结构和性质会发生改变，可以直接观察或通过仪器测量到明显的变化，据此可判断不同还原方法的效果。

2.1.1 光学特征

由于还原过程可以显著地提高氧化石墨的载流子浓度和迁移能力,这会显著增加对于入射光的反射,使薄膜呈现金属光泽,而 GO 膜则通常为棕色,如图 2.1(a)所示。光学透射和吸收测试可以直接用来观察 GO 还原前后的变化[1]。

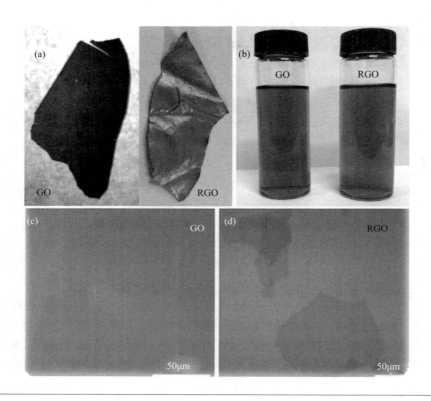

图 2.1 典型的光学图片
(a) GO 薄膜和 RGO 薄膜;(b) GO 溶液和 RGO 溶液;(c)、(d) 在 SiO_2/Si 基底上的 GO 和 RGO 薄片[1]

对于 GO 悬浮液而言,还原作用会带来明显可见的视觉特征变化[1,2]。如图 2.1(b)所示,将胶体状态的 GO 经化学还原(如肼)制备 RGO 时,黄棕色的 GO 悬浮液会被还原为黑色沉淀物。这是因为还原过程减少了 GO 上的极性基团,尤其是边缘区域极性降低,导致材料疏水性增强,以致出现团聚和沉淀。通过添加表面活性剂或者调节溶剂特性等方法可以继续保持 RGO 的胶体状态。

此外,如图 2.1(c) 和图 2.1(d) 所示,将 GO/RGO 片置于合适的基底(如 SiO_2/Si)上,利用光学显微镜可以在微观尺度上观察到还原前后的区别。由于 GO 片的绝缘特性,GO 片和基底之间只有微弱的光学衬度变化,几乎呈透明状态;

在GO片边缘较高衬度区域，则对应较高的厚度，通常是边缘折叠导致的[3]。而在还原之后，同样条件下RGO片和基底之间的对比度明显增强。此外，RGO在紫外可见吸收谱中可呈现更加显著的π-π*吸收峰，显示共轭结构部分恢复。

2.1.2 导电性

石墨烯具有良好的导电特性。纳米厚度的少层石墨烯片在室温下面电阻大约为 $400\Omega/\square$。Bae等人[4]报道将CVD法制备的石墨烯膜转移到透明基底上制备了4层石墨烯的透明导电薄膜（Transparent Conductive Film，简称TCF）。这种薄膜透明度约90%，面电阻大约为 $30\Omega/\square$。假设薄膜厚度是2nm，理论计算得到石墨烯薄膜电导率约为 1.6×10^5 S/cm，远远高于相同厚度的氧化铟锡（Indium Tin Oxide，简称ITO）或金属薄膜[5]。由于还原作用可以部分恢复石墨烯的高导电性，因此RGO的电导率可作为判断不同还原方法效果的一个直接标准。RGO的电导率可以用多种方式描述：RGO微片的电阻（$R_{s\text{-is}}$），RGO片组装的薄膜电阻（$R_{s\text{-f}}$），粉末电导（σ_p）和体电导（σ）。其中面电阻（R_s，$\Omega/\square$）可用来表征石墨烯微片或者薄膜电阻，且与厚度无关。

$R_{s\text{-is}}$ 的测量可以利用光学光刻或者电子光刻技术在RGO微片上组装微电极对，然后通过两探针或者四探针测试得到。Lopez等人[6]报道的RGO的 $R_{s\text{-is}}$ 约为 $14\text{k}\Omega/\square$（对应于360S/cm），比CVD石墨烯或者机械剥离的石墨烯高了大约两个数量级[7]。Su等人[8]测得的体积电导有1314S/cm。

对于石墨烯微片组装成的薄膜材料，研究者也经常用四探针方法测量膜表面的电阻（$R_{s\text{-f}}$）来描述其导电性。Zhao等人[1]通过氢碘酸化学还原制备了RGO基的TCF，膜厚大约10nm，在550nm的波长透明度为78%，测得了最低的 $R_{s\text{-f}}$ 约为 $0.84\text{k}\Omega/\square$，估算得到膜的体积电导率大约是1190S/cm。

此外，对于还原制备的石墨烯粉体材料，Stankovich等人[2,3,9]利用粉末电导来描述RGO的导电性。该技术先将RGO粉末压实成具有不同表观密度的薄片或者块体后用两探针或者四探针方法测试其体电阻，通过改变压力得到多个在不同压实密度下的电阻值，可进一步外推得到在石墨密度（约 2.2g/cm^3）下的理想粉末电导数值。

上述面电阻或者粉末电阻是判断氧化石墨还原程度的常用方法，但是需要在测试中注意探针与测试样品之间的接触问题以及粉末颗粒分布是否均匀引起的测量误差。

2.1.3 碳氧比

由于制备方法的差异，GO的化学组成可以是接近 $C_8O_2H_3$、$C_8O_4H_5$，也可以是非整数成分比；根据氧化程度不同，相应的碳氧原子比可在4:1～2:1之间变化或者超出这个范围[9,10]。还原之后，RGO的碳氧比可提高至12:1甚至更高[11,12]。测试碳氧比例可以直接判断还原过程中氧原子去除程度。

碳氧比可以通过元素分析（燃烧法）测量，也可通过 XPS 等元素谱学分析得到。燃烧法元素分析可以得到样品总成分的信息，而 XPS 主要是一种表面分析技术[11]，对于很多形式的样品两者数据基本相符，但也有一些情形两者相差较大。此外，XPS 可以提供更多关于还原过程中的化学结构变化信息，尤其是碳原子杂化形式及官能团种类与含量的一些信息。由于 sp^2 碳的 π 电子在很大程度上决定了石墨烯材料的光学和电学性质[13]，因此 sp^2 碳的含量对于研究结构与性能之间的关系具有重要参考意义。

XPS C1s 谱在还原前后的变化可清晰表明氧化程度的不同，四种形式的碳氧键可以较为清晰地进行分辨[14]：非氧化的碳环中的 C（约 284.6eV），C—O 键中的 C（约 286.0eV），羰基中的 C（约 287.8eV）以及羧基中的 C（O—C=O，约 289.0eV）。还原后，RGO 中这些含氧官能团对应的峰强大幅度降低；碳碳键对应的半峰宽也由于 sp^2 碳含量的提升而变得更窄。然而，由于 RGO 中含氧官能团的情况仍然比较复杂，利用 XPS 进行精确分析仍然存在挑战。

2.1.4 其他技术

除了上述指标和方法外，其他诸如拉曼光谱、SSNMR、TEM 以及 AFM 等也常被用来表征 GO 还原前后结构和性质的变化。这些测试可以提供关于 GO 和 RGO 更详细的信息，有助于理解还原过程的结构与成分变化，在大多数情况下，这些测试技术要从不同角度相互验证和配合才能较为清晰地解释还原过程。

2.2 还原方法与机理

2.2.1 热还原

（1）热退火

GO 可通过简单的热处理被还原，该过程通常被称作热退火还原。在石墨烯研究中，常通过快速加热（高达 2000℃/min）来剥离氧化石墨并同时实现还原以得到石墨烯[11,15~17]。一般认为，热还原剥离的机理是：GO 快速升温使含氧基团发生分解生成 CO、CO_2、H_2O 等气体，导致石墨烯片层间气体急速膨胀，从而产生巨大的压力，将堆叠的片层分开。依据哈梅克常数评估，仅 2.5MPa 的压力就足以将堆叠的 GO 分开，而基于气体状态方程进行估算，300℃ 可以产生 40MPa 气压，在 1000℃ 可产生 130MPa 气压[16]。

由于快速加热过程不仅可以剥离氧化石墨，还使得含氧基团在高温下发生分解从而去除部分含氧官能团，还原后的氧化石墨经常被称为热还原氧化石墨或者直接称为石墨烯。双重效应使热退火还原氧化石墨成为大量生产石墨烯的良好策略。

热退火这一过程也会产生小尺寸和起皱的石墨烯薄片[11]。在氧化石墨热退火过程中,氧化石墨片层上的含氧基团发生分解的同时也会带走片层上的碳原子,这会使石墨烯片撕裂成小片,并使片层发生扭曲,如图 2.2 所示。热剥离一个显著的影响是二氧化碳等气体的释放会引起石墨烯薄片的结构损伤[18],因此氧化石墨在热剥离过程中可能会有超过 30% 的质量损失,同时在整个片层留下晶格缺陷。缺陷的存在引入了散射中心,减少了弹道传递路径长度,不可避免地会对石墨烯的电子特性造成影响。因此,有研究发现热剥离得到石墨烯薄片的电导率平均值仅为 10~23S/cm[18],导电性比其他方法如机械剥离石墨或者 CVD 法制备的石墨烯低得多,表明该方法对碳平面结构的修复作用较为微弱。

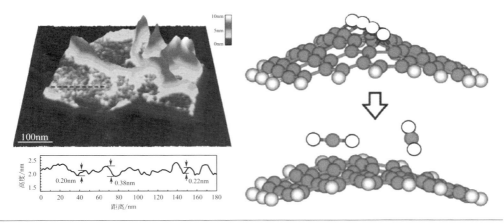

图 2.2 对单个石墨烯薄片进行 600nm×600nm AFM 扫描的 3D 图,显示了表面的褶皱和粗糙结构以及 GO 向石墨烯过渡的原子模型[11]

另一种方法是先在液相中剥离氧化石墨,这样可以获得大尺寸的 GO 微片[1],然后将其组装成宏观材料,如薄膜或粉末,再在惰性气氛或还原性气氛中退火可以制备得到石墨烯或者组装体。这样剥离再组装的过程改变了 GO 片层间的间距,从而也改变了热退火时气态压力的积聚与释放过程,因此,得到的石墨烯可具有不同的碳含量和结构完整性。

加热的条件,例如温度,显著影响还原效果[11,12,19~22]。Schniepp 等人[11] 发现当加热温度低于 500℃时,碳氧比低于 7,而当加热温度达到 750℃时,碳氧比高于 13。Li 等人[22] 监测了 GO 在不同气氛不同退火温度下化学结构的变化,表明需要高温才能达到良好的还原效果,XPS 谱图如图 2.3 所示。

Wang 等[23] 将 GO 薄膜在不同温度进行退火,发现在 500℃还原的石墨烯薄膜体电导率为 50S/cm,而在 700℃和 1000℃条件下还原的石墨烯薄膜体电导率分

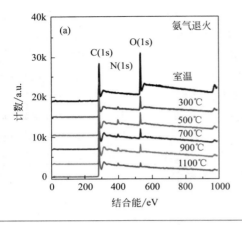
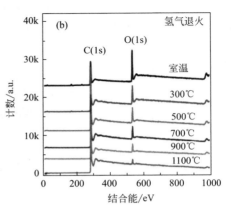

图 2.3 GO 经不同温度退火处理的 XPS 谱图
(a) 在 NH₃/Ar (10% NH₃) 气氛中；(b) 在 H₂ 气氛中[22]

别为 100S/cm 和 550S/cm。Pan 等[21] 使用电弧放电方法剥离氧化石墨制备石墨烯，电弧放电可以短时间内产生 2000℃的高温，得到的石墨烯膜电导率为 2000S/cm，元素分析测试表明这种剥离的石墨烯片碳氧比为 15~18。

除退火温度外，退火气氛对热退火还原氧化石墨也很重要。在高温下，氧气的蚀刻作用会急剧加强，所以以增加碳含量为目标时应在退火过程中应尽可能排除氧气。因此退火通常在真空或惰性及还原性气氛中进行。Li 等[22] 报道在低压氨气氛围下（NH₃/Ar，10% NH₃），可得到氮掺杂的还原氧化石墨。如图 2.3 所示，500℃退火可得到氮掺杂含量最高为 5%的石墨烯，在 NH₃ 氛围下退火得到的石墨烯比在 H₂ 氛围下退火得到的石墨烯具有更好的导电性，其具有明显的 n 型电子掺杂行为，有利于在光电材料等方面的应用。

此外，Lopez 等[6] 的研究结果表明，在高温（800℃）条件下将 RGO 暴露在碳源（如乙烯）氛围中，片层中存在的缺陷可以被部分修复。这个条件接近于 CVD 法制备单壁碳纳米管（Single-Walled Carbon Nanotube，简称 SWCNT）的条件。随着碳原子的沉积和结构修复，RGO 的电阻可以降低至 28.6kΩ/□（或者 350S/cm）。

综上所述，热退火可以高效还原氧化石墨，但也存在十分明显的缺点。首先，高温意味着能源消耗以及严格的处理条件；其次，如果对 GO 结构组装体（如 GO 薄膜）进行还原，那么加热速度必须足够慢，以避免破坏组装结构，但缓慢加热使热还原过程变得非常耗时；再者，在有些应用中需要将 GO 在基底上进行组装，低熔点基底（如玻璃和高分子聚合物）决定了热退火的温度不能过高，也就意味着较难彻底进行还原。

（2）微波还原和光照还原

区别于常规热源退火方式，一些其他加热手段也被用来实现热还原，包括微波辐射[24,25]以及光照辐照[26,27]等。

利用微波还原和剥离氧化石墨是一种既通用又简单的方法。Zhu 等[24]在空气氛围下，在微波炉中处理氧化石墨粉，1 分钟之内即可获得 RGO。Voiry 等[28]发展了此方法，用微波处理预还原的石墨烯薄膜，在 1～2s 之内可以得到电子迁移率大于 $1000m^2/(V·s)$ 的石墨烯，品质可以与物理剥离制备或 CVD 法制备的石墨烯相媲美。

Huang 等[26]报告了一种室温下非化学的闪光还原过程。这个过程利用照相机闪光灯瞬间触发光热加热发生脱氧反应。相机闪光灯在近距离内发出的光能（<2mm，约 $1J/cm^2$）可以提供 9 倍于加热 GO（厚度为约 $1\mu m$）至 100℃所需的热量，这些热量足以引发脱氧反应，并且表明闪光辐照可以导致更大程度的还原。由于快速脱气，GO 薄膜一般会膨胀数十倍，而膨胀薄膜的电导率则在 10S/cm 左右（使用最大的膜厚值进行计算）。利用光掩膜可以很容易地在柔性衬底上制作类似交叉数字电极阵列这样的导电图案，这有助于直接制造基于 RGO 薄膜的电子电路，如图 2.4 所示。

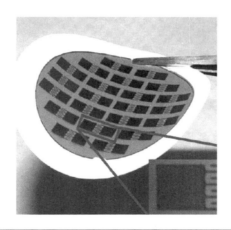

图 2.4 闪光还原制备带有图案的 RGO 薄膜[26]

Zhang 等[27]人提出用飞秒激光还原氧化石墨。实验中使用的聚焦激光束（激光脉冲的中心波长为 790nm，脉冲宽度为 120fs，重复频率为 80MHz，聚焦物镜为 100 倍）功率密度高于氙闪光灯，线宽范围在 10^{-1}～$10\mu m$。因此，飞秒激光还原可以制备更高电导率（256S/cm）的 RGO 膜，并且可以用预定程序的激光扫描直接在 RGO 膜上绘制图案。利用得到的石墨烯微电路测量的电流电压曲线显示了典型的线性关系,表明了该微电路具有稳定的导电率和良好的接触。进一步发展飞秒激光还原技术可以拓展石墨烯基材料在电子微器件领域应用。

2.2.2 光催化还原

除了光的热还原作用,GO 也可以在光催化剂(如纳米 TiO_2)作用下发生光-化学还原反应。Graeme 等人[29] 报道了 GO 胶体在 TiO_2 粒子和紫外线照射的条件下发生的还原反应,胶体颜色从淡褐色转变到暗褐色再到黑色。颜色的变化表明了碳-碳双键共轭结构得到一定程度上的修复,其变化过程与作用机制如图 2.5 所示。Kamat 等[30] 还对半导体 TiO_2 的光催化还原 GO 过程进行了研究。在紫外线照射下,TiO_2 颗粒表面发生了电荷分离形成空穴和电子。空穴在乙醇存在的条件下形成自由基,从而使电子在 TiO_2 粒子中富集。TiO_2 粒子中富集的电子对 GO 发生还原作用。这种光-化学还原方式也可以用来还原其他碳材料,例如在碳纳米管和富勒烯相关领域的类似研究报道[31,32]。

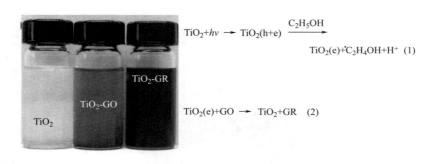

图 2.5 TiO_2-GO 乙醇溶液在 UV 辐照前后的颜色变化以及 TiO_2 催化 GO 还原的机理示意[29]

在上述的还原过程中,GO 上的羧基可通过电荷转移与 TiO_2 表面上的羟基相互作用,形成 GO 和 TiO_2 的纳米复合物,并且这种结构在还原后依然能够保持。经过光催化还原后的 RGO 可以作为一个电流收集器[33],用在器件上促进形成电子/空穴对,如光催化装置和染料敏化太阳能电池[34,35]。类似的,一些其他的光催化活性的材料,如 ZnO 和 $BiVO_4$,也被报道可以促进 GO 的还原反应[36,37]。

2.2.3 溶剂热还原

除了在干燥气氛下热退火还原等途径,还可以在溶剂中直接加热 GO 悬浮液还原,即溶剂热还原法[38~40]。溶剂热还原过程通常在密封容器中进行,加热时容器内的压力增加,悬浮液温度可以升温至溶剂沸点以上[41]。例如,在水热处理过程中,超临界水可以起到还原剂的作用。此外,随着压力和温度的变化,水的物理化学性质也随之改变，可以催化极性（离子）键在水中发生裂解反应。水热处理的方法已经被用于利用大分子含碳化合物制备均一性良好的碳纳米球[42,43]、碳纳米管[44]。Zhou 等[38] 通过水热法处理 GO 溶液提出了一种"只含水"的溶剂热处理方法，发现超临界水不仅部分去除了 GO 的官能团，而且修复了碳晶格中的芳香结构。研究发现，在碱性溶液（pH=11）水热还原可以制备稳定均一的 RGO 溶液，而在酸性溶液（pH=3）中水热会导致 RGO 发生聚集，即使在浓氨水中也无法再分散。作者认为，该还原过程类似于质子催化酒精脱水过程，其中质子是形成羟基的来源。

  有机溶剂具有比水更高的沸点。Wang 等[39] 人报道了用 DMF 作为溶剂进行 GO 的溶剂热还原反应的方法。该试验中，研究人员在 DMF 中加入少量的肼作为还原剂，经 180℃热处理 12 小时后，利用俄歇能谱检测到 RGO 的 C/O 比例达到 14.3，远高于常压下肼的还原程度。得到的 RGO 膜表面电阻在 $10^5 \sim 10^6 \Omega/\square$ 范围，被认为与肼还原引起的氮掺杂相关。Sergey 等[40] 提出了一种用 NMP（N-甲基吡咯烷酮）作为溶剂改进的溶剂热还原方法。该方法无需在密闭条件下进行热处理，可以在低于 NMP 沸点的 200℃下进行。作者认为还原机理是热效应以及高温下 NMP 具有清除氧的性质相结合引起的脱氧反应。以该方法制备的 RGO 膜，电导率为 3.74S/cm，比利用肼还原的产物（82.8S/cm）低了一个数量级。这种溶剂热还原的 GO 的碳/氧比例仅为 5.15，远低于以上其他溶剂热处理的结果。

  溶剂热还原方法可在还原的同时产生稳定分散的 RGO，这对于 RGO 的应用具有一定的价值。

2.2.4 电化学还原

  另一种还原 GO 的方法是电化学法去除含氧基团[45~47]。例如，在室温下以水溶液作缓冲液，可在电化学电池中进行 GO 的电化学还原。电化学还原反应主要依靠 GO 与电极之间的电子交换，通常不需要特殊的化学试剂，可以避免使用危险还原剂如肼，并消除了副产物。

  Ramesha 和 Sampath[47] 将 GO 沉积在基底上（玻璃、塑料、ITO 等）作为工作电极，使用惰性电极作为对电极，以 $0.1M\ KNO_3$ 为电解液，在 $0\sim0.1V$ 电压范围（相对于饱和甘汞电极）对 GO 改性的电极进行循环伏安扫描，伏安曲线测试发现 GO 还原反应开始于约 0.6V，并在约 0.87V 达到了最大值。

  Zhou 等[45] 人报道了使用电化学方法的较好还原效果；对产物 RGO 进行元素分析显示，C/O 比例为 23.9，RGO 薄膜的电导率为约 85S/cm。研究发现，控制缓冲溶液的 pH 是控制还原反应的关键因素，较低的 pH 有利于 GO 的还原反应，并提出了 H^+ 参与的反应机制。

2.2.5 化学试剂还原

  化学试剂还原是利用具有还原性的化学试剂还原 GO 的方法。

(1) 硼氢化物

硼氢化钠（Sodium Borohydride，$NaBH_4$）是一种常见的硼氢化物，也是合成化学中常用的还原剂。这种盐具有四面体的 BH_4^-，易溶于水和乙醇介质。在缺电试剂（如羰基官能团）存在的情况下，硼氢化物会进行氢负离子转移反应生成氧离子和缺电子的 BH_3 分子，随后 BH_3 分子和氧负离子的稳定化作用会使硼氢化物复原；理想情况下硼氢化物作为氢负离子转移剂，直到所有的 B-H 键都被耗尽。BH_4^- 还原羰基的能力也受到羰基类型的限制，简单的羰基化合物，如醛和酮，可以由 $NaBH_4$ 还原成醇类，而酯或酰胺等活性较低的羰基化合物是不可被还原的。

Muszynski 等[48]人为制备吸附纳米金粒子的十八胺修饰石墨烯，报道了 $NaBH_4$ 作为 GO 还原剂的研究结果。同时，Si 和 Samulski[49] 通过一个三步反应制备磺化石墨烯，在水中和有机溶剂中均有良好的分散性，其中第一步反应就是利用 $NaBH_4$ 作为还原剂还原氧化石墨，这种预还原的磺化石墨烯的电导率为 17S/m。

Gao 等[50]人以 $NaBH_4$ 作为还原剂，采用类似的方法制备还原程度较高的 GO。Shin 等[51] 开展了 $NaBH_4$ 浓度对还原后石墨烯导电性能影响的研究。通过将 GO 浸入 150mM 的 $NaBH_4$ 溶液中，制备了电导率为 45S/m、碳氧比为 8.6 的石墨烯，其电阻低于肼还原的石墨烯，这可能和 $NaBH_4$ 还原过程中引入杂原子有关。

另一方面，出于调控 GO 含氧官能团种类的目的，研究人员对硼氢化钠、三乙酰基三乙酰氧基硼氢化钠 [$NaBH(OAc)_3$]、氰基硼氢化钠（$NaBH_3CN$）的还原能力进行了研究[52]。这些反应均在甲醇溶剂中进行，制备的石墨烯的碳氧比（C/O）范围在 2.2～2.5。电化学阻抗谱学研究表明 $NaBH_4^-$、$NaBH_3(CN)^-$ 以及 $NaBH(OAc)_3$-Graphene 的电荷转移电阻分别是 $1.64k\Omega$、$1.72k\Omega$ 及 $4.92k\Omega$（相对比而言，氧化石墨：$2.27k\Omega$，无定型碳：$0.6k\Omega$）。

氨硼烷是一种温和的还原剂，具有和硼氢化钠类似的反应活性。Pham 等[53] 使用氨硼烷分别在水以及有机溶剂中还原氧化石墨烯，B 和 N 是以 BC_2 以及吡咯型 N 的形式掺杂于石墨烯中，在有机溶剂中还原，制备的石墨烯具有更高的 B 和 N 含量，并表现出更好的超级电容器性能。

(2) 氢铝化物

在合成化学中，氢化铝锂是最强的还原剂之一，其还原能力超过硼氢化盐，因此更多的含氧基团如羧酸、酯可以被还原成相应的羟基。相对于硼氢化钠和肼[54]，使用氢化铝锂还原的石墨烯碳氧比可达到 12（$NaBH_4$：9.5，N_2H_4：11.5）。氢化铝锂对于 GO 的还原程度可以通过检测石墨烯上的亚甲基（—CH_2—）来表征。一般认为，这种亚甲基是由不饱和羰基烯烃加氢得到。此外，氢化铝锂还原得到的石墨烯在水溶液中的分散稳定性较低，这可能和还原程度较高有关。交叉金电极的电导率测试表明该石墨烯具有较低的电导率，仅为 2.9 S/m（$NaBH_4$：62.4S/m，N_2H_4：46.6S/m），这可能和亚甲基的存在有关。

（3）氢卤酸

氢卤酸如 HI、HBr 及 HCl 在合成化学中经常被用于亲电加成和亲核取代反应。卤化物本身是亲核的，亲核性按照 Cl^-、Br^-、I^- 顺序依次递增，通过单分子亲核取代反应（SN_1）或者双分子亲核取代反应（SN_2）与反应物进行反应。卤化氢也能够打开环氧基团。这些反应使氢卤酸能够很好地去除 GO 上丰富的羟基和环氧基团。

Moon 等[55]人根据早前的一篇报道，将多环醌和酚与氢碘酸和醋酸混合，可以得到相应的芳烃。当 GO 在两种不同相里进行还原时，液相反应比气相反应更有效率。液相反应制备的石墨烯碳氧比为 11.5，电导率为 30400S/m。燃烧离子色谱分析结果表明得到的石墨烯中不存在残留的 I^- 或者 I_2。

该研究小组更进一步的研究表明，氢碘酸和三氟乙酸的混合物可以在零度以下（-10℃）还原氧化石墨烯[56,57]。三氟乙酸的酸性是乙酸的 100000 倍，且具有更低的冰点-15.4℃。该混合物还原石墨烯具有高达 12.5 的碳氧比，表面电阻为 2Ω/□。Pei 等[57]进行了类似的研究，发现仅用氢碘酸作为还原剂，在 100℃下还原氧化石墨烯，得到的石墨烯具有 12 的碳氧比，电导率为 298S/cm。X 射线光电子能谱测试结果表明，即使在 400℃下经过 2 小时的退火处理，依然有残留的 I^- 和 I_2。该反应的机理和 Lee 等人提出的观点类似，碘负离子进攻环氧基团进行开环，然后消除结合能较弱的碳碘键。

Chen 等[58]利用氢溴酸还原氧化石墨烯，制备的石墨烯具有 0.023S/m 的电导率，残留的溴化物可能是造成电导率低的原因。此外，Chua 等[59]的研究发现用叔丁醇钾处理溴化石墨烯会发生脱溴化氢反应，这会提高石墨烯的导电性。得到的脱氢溴化石墨烯导电率为 51S/m，溴化石墨烯电导率为 39S/m，用肼还原石墨烯电导率则为 34S/m。

（4）含硫还原剂

① 二氧化硫脲 长期以来，二氧化硫脲在纺织、造纸、照相和皮革加工行业被用作强还原剂。据报道，二氧化硫脲是一种优质的环氧酮除氧剂，可还原酮类、芳香族硝基、偶氮、氧化偶氮、肼和有机硫化合物。另外，二氧化硫脲价格低廉，并且还原反应仅产生普通工业废物，如尿素和亚硫酸钠，这也为石墨烯的规模化生产提供了一个机会。

二氧化硫脲的上述性质使其可以作为 GO 的还原剂[60]。根据对石墨烯还原程度和电化学性能的需求，还原反应时间可以从 1~20 小时之间调控。制备的石墨烯的碳氧比为 14.5，并且能够在水中分散。电化学阻抗测试表明该种石墨烯电荷转移电阻为 0.11kΩ（相比而言，氧化石墨：5.43kΩ，玻碳：0.76kΩ）。Wang 等[61]

报道硫脲在碱性（如 NaOH）条件下具有相当的还原能力。硫脲还原得到的石墨烯，碳氧比为 5.89，电导率为 3205S/m。Ma 等[62]报道在 NH_3 代替 NaOH 提供的碱性条件下，以二氧化硫脲作为还原剂，制备的石墨烯碳氧比为 6，电导率为 290S/m。

② 乙硫醇-氯化铝　上述部分还原剂，如硼氢化钠和锂铝化物，只能将含氧的基团还原至羟基，但没有去除。受到 Node 等的启发，许多研究者研究选择性地移除羟基的方法，基于单电子转移方法，将烷氧基、过氧基、羟基、烷硫基从芳香族化合物上脱除[63]。一些研究者利用乙硫醇-氯化铝复合物还原氧化石墨烯[64]。用乙硫醇-氯化铝复合物还原石墨烯制备的石墨烯具有 4.71 的碳氧比，在水中具有稳定的分散性，这可能和未被还原的羧基官能团有关。电化学阻抗测试表明该石墨烯具有低的电荷转移电阻，为 0.09kΩ（相比而言，氧化石墨：4.56kΩ，玻碳：0.55kΩ）。

③ 劳森试剂　劳森试剂 [2,4-双（对甲氧苯基）-1,3-二硫—二磷杂环丁烷-2,4 硫化物] 具有交替的硫和磷组成的四元环结构，常用于将酰氨基或酯基中的羰基转换为硫代羰基。劳森试剂的反应活性取决于其活性中间体——二硫代磷酸物，硫代反应生成稳定的 P=O 键，类似于维蒂希反应。除此之外，劳森试剂也会使羟基发生硫醇化[65]。Liu 等[66]报道了使用劳森试剂还原氧化石墨烯，通过将 GO 片或膜浸入劳森试剂中，制备的石墨烯电导率为 4760S/m。XPS 数据显示这样制备的石墨烯的硫含量为 1.03%（原子）[氧化石墨为 0.94%（原子）]。该研究小组认为，活性二硫代磷酸物亲核的硫原子可能与 GO 上的含氧基团形成强烈的相互作用，从而导致环氧基团的开环；二硫代磷酸物也可直接亲核进攻消除羟基。更重要的是，该研究小组认为相邻的羰基通过烯化作用修复了氧化石墨上的 sp^2 共轭结构，从而解释了制备的石墨烯具有较高的电导率。将该法制备的石墨烯进一步在 300℃ 退火，电导率可提高至 30900S/m。相比而言，直接将氧化石墨在 300℃ 退火，电导率仅为 2260S/m。

④ 过硫酸钠　过硫酸钠是一种纺织行业传统的还原剂。Zhou 等[67]将过硫酸钠用于氧化石墨的还原。该研究小组将过硫酸钠加入碱性的氧化石墨悬浮液中反应 15 分钟后，制备的石墨烯电导率为 1377S/m（氧化石墨为 $1×10^{-4}$S/m）。其中，氢氧化钠在反应中作为催化剂和稳定剂。该研究小组认为，$Na_2S_2O_4$ 在碱性溶液中具有低电极电位（E_0（$SO_3^{2-}/S_2O_4^{2-}$）=-1.12V）促进了还原反应。因此，$S_2O_4^{2-}$ 亲核性攻击 GO 的环氧基团或羟基，然后释放 H_2O 分子，形成中间产物。随后，中间产物热消除导致石墨烯的形成，而 $S_2O_4^{2-}$ 被氧化为 SO_3^{2-}。

⑤ 硫脲　Liu 等[68]人试图利用硫脲大规模制备石墨烯。实验发现，在 95℃ 下经 8h 反应后制备的石墨烯碳氧比为 5.6，电导率为 635S/m，产物在乙醇和 DMF 中具有良好的分散性。该研究小组认为 C=S 键活化了二胺，二胺与 GO 表

面的环氧基团发生环化-脱除反应；二胺与 GO 边缘的羰基发生接枝反应，然后进行脱氧反应；二胺与相邻羟基发生氢键作用，然后脱氧。少量的接枝产物的静电作用可能是保持石墨烯在有机溶剂中良好分散的主要原因。

⑥ 噻吩　Some 等[69]人利用噻吩还原氧化石墨，噻吩作为一种双功能还原剂，同时体现了还原和修复的作用。噻吩和 GO 在 80℃下反应 24 小时制备的石墨烯碳氧比为 10.9（N_2H_4-graphene：14），面电阻为 59Ω/□（N_2H_4-graphene：26Ω/□）。反应过程中噻吩发生氧化和聚合，释放电子，GO 得到电子被还原成 RGO，反应过程释放的 SO_2 气体与噻吩聚合物形成 p-共轭的聚烯吸附在石墨烯表面，随后经 800℃热退火将吸附的聚烯去除。退火后的噻吩还原石墨烯的碳氧比可提高至 16.8，面电阻为 9Ω/□。此外，噻吩还原得到的石墨烯经退火后拉曼 I_D/I_G 比值为 0.41。相比而言，直接退火后得到的石墨烯 I_D/I_G 比值为 0.83（氧化石墨 I_D/I_G：0.91），表明了聚烯在还原过程具有一定的修复能力。

（5）含氮还原剂

① 肼　水合肼作为一种常见的抗氧化剂，工业领域被广泛使用。它在清除氧的同时分解为氮和水，但同时它也是一种剧毒物质。Stankovich 等[2]利用水合肼作为还原剂制备了碳氧比为 10.3、电导率为 2420S/m 的石墨烯（石墨的电导率为 2500S/m）。由于含氧基团的减少，获得的石墨烯具有较高的疏水性。

Ruoff 小组提出了环氧基团的还原过程，研究假定水合肼通过亲核反应进攻环氧基团生成醇肼类物质，然后释放一个水分子，向氮杂环丙烷类物质转变，最后经热消除二酰亚胺生成双键。基于密度泛函理论的计算表明，水合肼夺取环氧基团的 H，促使了环氧基团开环[70]。此外，由于水合肼和 GO 上的酸酐或醌发生反应，生成酰肼和腙，从而解释了残余氮原子的存在。虽然这一还原过程的机理解释仍存有疑问，但是该研究揭示了石墨烯边缘五元吡唑环及吡唑环的生成。研究认为，化学取代反应优先在边缘处发生，二酮类基团最丰富。

另一方面，在 DMF/H_2O 混合介质中用水合肼还原氧化石墨可以得到石墨烯胶体[71]，无须表面活性剂或稳定剂，就可在溶剂中可以保持较好的分散稳定性；但在丙酮、四氢呋喃、乙醚、甲苯、3,3′-二氯联苯胺（3,3′-Dichlorobenzidine，简称 DCB）、二甲基亚砜（Dimethyl Sulfoxide，简称 DMSO）、乙醇、N-甲基吡咯烷酮（N-Methyl pyrrolidone，简称 NMP）和乙腈中都没有获得稳定的胶体。制备的石墨烯电导率为 1700S/m（在 150℃下干燥后电导率为 16000S/m）。此后，Viet 等[72]使用苯肼还原氧化石墨烯，希望改善石墨烯的分散稳定性。所得的石墨烯在二甲基乙酰胺（Dimethylacetamide，简称 DMAC）、DMF、NMP 和碳酸亚丙酯中可稳定至少 2 个月。这种稳定性主要是由于苯肼的空间效应阻碍了石墨烯的团聚。该石墨烯电导率为 4700S/m（在 150℃下干燥后电导率为 21000S/m）。

② 羟胺　羟胺是一种常见的功能基团（如：肟类和异羟肟酸等）的前驱体，广泛应用于尼龙 6 的生产和金属离子（如 Cu^{2+}、Pd^{2+}、Ag^+ 和 Au^{3+}）的还原。

此外，羟胺具有自由基清除剂和过氧化物分解剂的作用。由于其抗氧化特性及与水反应弱的性质，因此被用作 GO 的还原剂。Zhou 等[73] 利用盐酸羟胺和氨原位反应生成羟胺来还原氧化石墨烯，制备的石墨烯碳氧比为 9.7，电导率为 1122S/m。该法制备的石墨烯在水中可形成稳定的悬浮液，这可能是由于反应物具有碱性导致石墨烯表面带上了相互排斥的负电荷造成的。该研究团队还提出了一系列羟胺还原氧化石墨烯的机理，如图 2.6 所示。其中环氧相关的还原路径是，羟胺首先引起环氧开环，随后通过质子转移和水分子的消除获得 N-羟甲基吡啶中间体。在去除羟基的过程中也得到了类似的中间产物。最后，释放不稳定的氮氧化物和水分子的同时，每一个 N-羟甲基吡啶中间体被转化成一个共轭乙烯基。作者认为羟胺不太可能还原羰基。

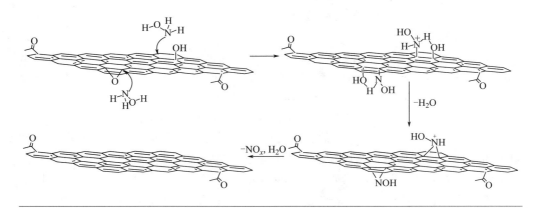

图 2.6 羟胺还原羟基与环氧化物的机理示意[73]

Mao 等[74] 人道了用盐酸羟胺还原氧化石墨烯，制备的石墨烯 C/O 比为 1.5，经 200℃ 退火后，C/O 比提高至 2.5，高分辨率 XPS 分析显示了该石墨烯材料上存在氮掺杂。该石墨烯在 DMF 中保持分散稳定长达一周。另外，通过直流电测得的石墨烯的电阻约为 $10^5\Omega$（GO 约为 $10^{10}\Omega$），经退火处理后进一步提升为约 $10^2\Omega$。该小组提出，在 GO 的反应中，羟胺中 N^- 释放电子，再结合反应混合物中存在的 H^+，以生成水的形式脱除含氧基团。如下式所示：

$$GO + 2e^- + 2H^+ \longrightarrow RGO + H_2O$$

该法制备的石墨烯表现出 p 型半导体特性。当作为气体传感器测试时，所制备的石墨烯对稀释于干燥空气中的 NO_2（0.01%）和 NH_3（1%）表现出快速响应和高灵敏度的特性。

③ 吡咯 吡咯是弱碱，是合成天然产物的先驱体。Sakura 和 Nagasaki[75] 利用吡咯衍生物作为还原剂来制备金胶体。此后，Amarnath 等[76] 研究发现，吡咯

作为还原剂和稳定剂，得到的石墨烯C/O可达7.7，同时在乙醇、异丙醇、DMF、DMSO、NMP、THF和丙酮等有机溶剂中显示出良好的分散稳定性。这种稳定性被认为是由吡咯吸附在石墨烯表面上的氧化产物赋予的。至于吡咯的还原机理，推测是电子从吡咯转移到GO，而吡咯本身被氧化形成低聚物。

④ 苄胺 Liu等[77]人将苄胺用作GO的还原剂，苄胺也可用作所得石墨烯材料的稳定剂。用苄胺处理GO1.5小时，得到石墨烯的C/O比为4.7；当反应时间延长至2小时，C/O比增加至10.2，所得到的石墨烯的电阻为$5\times10^8\Omega$，且在水溶液中表现出良好的分散稳定性。吸附在石墨烯苄胺分子的极性氨基和带负电的羧基的存在与这种分散稳定性有关，苄胺的-NH_2和GO直接进行氧化—还原反应。

⑤ 对苯二胺 Chen等[78]人开发了一种使用对苯二胺（para-Phenylene Diamine，简称PPD）的还原方法，其中对苯二胺既作为还原剂，又起到稳定剂的作用。通过这种方法制备的石墨烯因为吸附了氧化PPD（含有-N^+残留基团）而带正电荷，抑制了石墨烯片之间π-π相互作用和聚集。因此，该石墨烯在乙醇、乙二醇和NMP中的分散稳定性增强（DMF仅有短期稳定性）。此外，所得的石墨烯膜平面电导率为15000S/m。但是反应机理尚不清楚。

⑥ 乙二胺 Che等[79]人提出了另一种用二胺化合物还原氧化石墨烯的方法，特别是乙二胺。研究人员旨在用乙二胺使GO功能化，但意外地实现了GO的还原。基于XPS分析发现，制备的石墨烯含有N掺杂，C/O比为7.87，测量的电导率为220S/m（氧化石墨：0.02S/m，肼还原：1300S/m）；该石墨烯在DMF中可形成稳定的分散体，这可能是由于石墨烯与DMF的相容性增加以及石墨烯表面上存在酰胺基团。以正丁胺还原氧化石墨作为对照实验，发现还原程度较低，因此认为乙二胺中二氨基团的存在导致了氧化石墨的还原。该研究认为环化脱除是占主导地位的还原机理，乙二胺和环氧基团发生开环反应，相邻的羟基发生环化生成哌嗪，随后被除去以释放环结构的张力。除此之外，该团队还提出了乙二胺和相邻的羟基反应生成羟胺，然后脱除的反应过程。

⑦ 尿素 尿素在合成化学中具有悠久历史。Lei等[80]将其引入作为氧化石墨的还原剂。在碱性条件下，尿素还原氧化石墨烯制备得到的石墨烯在水、乙二醇和DMF中表现出长达36小时的分散稳定性。同时发现，乙醇和2-丙醇不适合用于制备石墨烯的分散液，氯仿完全不能分散石墨烯。尿素还原的石墨烯C/O比为4.5，电导率为43S/m（氧化石墨：0.0012S/m）。经过800℃下的退火后，石墨烯的C/O比为19.7，电导率为4520S/m。高分辨率XPS测试表明，该石墨烯具有1.2%（原子）的氮含量，这表明存在吡啶等含氮部分，但该还原过程的机理尚不清楚。

⑧ 二甲基酮肟 二甲基酮肟是一种锅炉水中常用的除氧剂。Su等[81]人将其应用于GO的还原。该研究利用二甲基酮肟处理GO的碱性水悬浮液得到石墨烯，其C/O比为6.53，电导率为100S/m（氧化石墨：<0.01S/m）。得到的石墨烯在水溶液中表现出优异的分散稳定性，长达6个月。除此之外，高分辨率XPS测试

表明该石墨烯具有1.2%（原子）的氮掺杂含量，包括吡啶、吡咯和石墨化氮。虽然确切的机理仍然有疑问，但Su等[81]人认为原位生成的羟胺与丙酮是活性还原成分，丙酮是二甲基酮肟在碱性条件下发生水解原位生成的。

（6）含氧还原剂

① 醇　简单的醇，即MeOH、EtOH、iPrOH和BnOH，对氧化石墨具有还原能力。Dreyer等[82]的研究证明，在这些醇中，BnOH具有最高的还原程度，制备的石墨烯C/O比为30，电导率为4600S/m。此外，还原后的石墨烯悬浮液中存在苯甲醛和苯甲酸，表明BnOH直接参与了还原过程。

② L-抗坏血酸　抗坏血酸，也称维生素C，是一种人体常见的必需营养素，具有抗氧化性。这一特性启发了Gao等的研究[83]，将其作为还原剂与L-色氨酸一起用于GO的还原。在这项工作中，L-抗坏血酸作为还原剂，而L-色氨酸通过π-π相互作用吸附在石墨烯片上来稳定生成石墨烯。L-色氨酸的稳定作用与剩余带负电的羧酸残余物发挥协同作用，在石墨烯片之间提供静电斥力。得到石墨烯的电导率为14.1S/m（相比而言，GO：5.72×10^{-6}S/m）。

研究认为，还原通过两步进行，如图2.7所示。首先发生SN2亲核反应，然后进行热消除反应。由于抗坏血酸五元环上的吸电子基团增加了β和γ-羟基基团的酸性，导致两个质子解离形成抗坏血酸氧阴离子（HOAO⁻）。然后，阴离子进攻GO的环氧基团及二醇基团，进行背面SN2亲核反应形成中间体。最后，中间体被

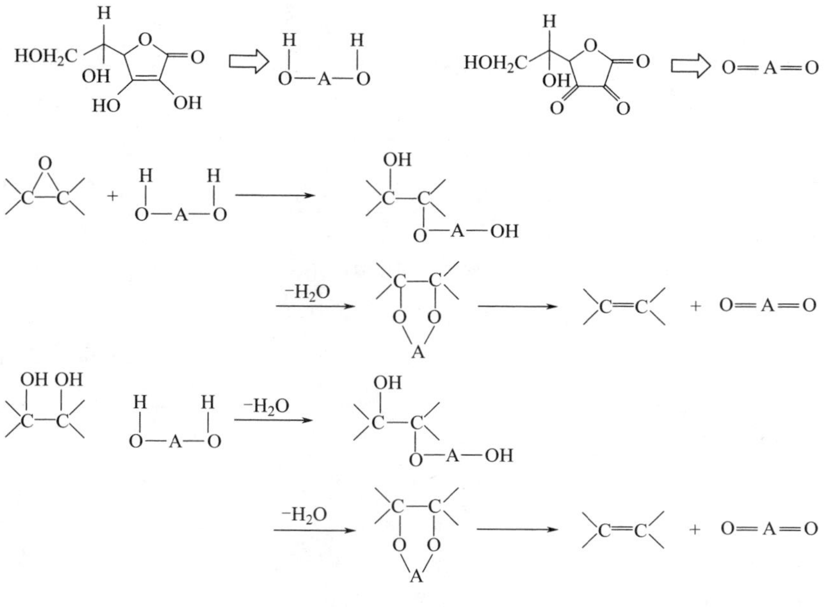

图2.7　L-抗坏血酸还原环氧基团和二羟基的机理示意[83]

热除去，抗坏血酸氧化为脱氢抗坏血酸，GO 转变为石墨烯。另外，Zhang 等[84]人报道了仅使用 L-抗坏血酸还原氧化石墨烯制备石墨烯的研究结果，其电导率为 800S/m。整个过程耗时 48 小时，而更高浓度的 L-抗坏血酸能缩短反应时间。得到的石墨烯在水中具有良好的分散稳定性。研究认为，脱氢抗坏血酸的分解产生草酸和古洛糖酸，这些酸可与石墨烯表面上残留的含氧官能团形成氢键，防止石墨烯片层间的堆积和 π-π 相互作用。

③ 糖类 糖一般分为单糖、二糖、寡糖和多糖，它们的化学结构中都有含氧官能团。单糖（如葡萄糖）和果糖是常见的还原糖，由于存在游离的羰基和羟基，所以能够还原 Tollen's 试剂或 Fehling's 溶液。这种通过氧化还原途径体现的还原能力高度依赖于糖类形成开链结构的能力，在非还原糖（如蔗糖）中通常观察不到。

Zhu 等[85]人在氨水溶液中使用葡萄糖、果糖和蔗糖用来还原氧化石墨烯。氨的存在对于还原程度非常关键，因为它有助于协同加快反应速率和脱氧过程。然后，醛糖酸进一步转化为内酯。富含羟基和羧基的醛糖酸和内酯的混合物可以与石墨烯表面上残余的含氧官能团形成氢键，以防止石墨烯片通过 π-π 作用堆叠。另一方面，由于果糖在碱性条件下能够经历酮-烯醇互变异构，因此果糖也表现出对 GO 的还原能力。至于蔗糖，尽管是非还原糖，但它在碱性环境中水解后易于分解为果糖和葡萄糖，也可以实现 GO 的还原。但是，其还原能力低于葡萄糖和果糖。此外，由葡萄糖处理获得的石墨烯体现出对多巴胺、肾上腺素和去甲肾上腺素的良好电催化活性。

在另一项工作中，Kim 等[86]人在氨水溶液中用葡聚糖（多糖）还原氧化石墨。类似于如前所述观察到的氨协同效应，氨的加入将还原反应从 3 天缩短至 3 小时。该反应的机理与葡萄糖还原类似。获得的石墨烯电导率为 1.1S/m（在 Ar 气氛中 500℃下进一步退火后电导率为 10000S/m）。

（7）金属酸还原剂

近年来，金属和酸混合作为还原剂在石墨烯领域受到了广泛的关注，研究发现混合物具有快速和高效的还原能力。目前，该还原方法的作用机制存在两种机制的探讨：a. 金属和 GO 之间的快速电子转移；b. 产生活性氢实现还原。

Fan 等[87]报道了利用混合试剂还原 GO 的研究结果。发现使用铝粉（10μm）和盐酸混合物在 30 分钟内还原氧化石墨烯，可得到 C/O 为 18.6 的还原氧化石墨烯，其电导率为 2100S/m。在无盐酸存在的条件下，铝无法还原 GO。该研究认为在此过程中盐酸将铝表面的致密钝化层溶解后，新生的铝与 GO 紧密接触，提高了电子转移的效率同时能产生活性氢将 GO 还原。在该团队的后续工作中，以铁粉（10μm 粒径）代替铝作为还原剂[88]，经 6 小时还原处理后获得的石墨烯样品 C/O 为 7.9，测得的电导率为 2300S/m，拉曼 I_D/I_G 为 0.32（相比而言，氧化石墨：0.86）。这说明 Fe 的存在使石墨烯的 sp^2 网络结构得到修复。研究发现石墨烯氧化物薄片包裹铁颗粒，猜测是由于铁表面上的 Fe^{2+} 和带负电荷的 GO 之间的静电引力造成的。紧密包裹促进了从 Fe/Fe^{2+} 到带负电的 GO 片材的快速电子传递，如下式所示。同样地，在无盐酸的条件下未观察到 GO 的还原。

$$GO + aH^+ + be^- \longrightarrow Graphene + cH_2O \qquad (2.2)$$

Mei 和 Ouyang 等[89]人研究了锌对于 GO 的还原，发现在超声条件下盐酸中的锌粉能够在 1 分钟内还原氧化石墨烯。当锌与 GO 的重量比为 2 时，得到的石墨烯的 C/O 比为 33.5，测得其电导率为 15000S/m。他们认为，这种自发的还原过程主要是由于 Zn^{2+}/Zn (0.76V) 和 GO 的标准还原电位（pH=4，0.40V）差值较大所驱动的。Ramendra 等[90]人报道了用固体锌屑和硫酸经 2 小时反应还原氧化石墨烯的研究，得到的石墨烯 C/O 比为 21.2，测得其电导率为 3416S/m。他们提出反应机理认为：羰基和环氧基团的还原通过一系列形成羟基的步骤，最终脱水得到烯烃，$ZnSO_4$ 的存在催化了脱水过程。羧基则是在氢质子催化下发生脱羧反应。

Barman 等[91]人用 2M 盐酸混合镁还原氧化石墨烯，观察到新生的氢是还原氧化石墨烯的主要还原剂，得到石墨烯的 C/O 为 3.9，电导率 10S/m。该研究用 Ag 基底的传感器检测到新生氢的转变，结果表明在酸性介质中镁比锌可产生更多的新生氢。

（8）金属碱性还原

铝和锌具有在碱性和酸性环境中的反应性，因此也可用于碱性介质中还原氧化石墨烯。

Liu 等[92]人在氨水存在条件下，利用锌粉还原氧化石墨烯，10 分钟内即可得到还原氧化石墨烯。推测可能的还原机理是"Zn-GO 原电池"，其中锌作为阳极，GO 作为阴极，溶解的氨作为电解质，如反应式所示[92]：

$$\begin{aligned} &Zn - 2e^- \longrightarrow Zn^{2+} \\ &NH_4^+ + e^- \longrightarrow NH_3 + H \\ &Zn^{2+} + 4NH_3 \longrightarrow [Zn(NH_3)_4]^{2+} \\ &GO + 2H \longrightarrow Graphene + H_2O \end{aligned} \qquad (2.3)$$

Feng 等[93]人使用钠-氨（Na-NH_3）还原氧化石墨烯。该研究将金属钠加入含有 GO 的液氨中，并将其于干冰-丙酮条件下保持 30 分钟。得到的石墨烯 C/O 比为 16.61，测得的薄膜面电阻为 350Ω/□。钠金属在液氨中的溶解提供了钠阳离子和与氨溶剂化的电子。溶剂化电子可能裂解碳-氧键，在 GO 表面形成碳自由基或自由基阴离子，去除环氧化物、羟基、羰基以及羧基。溶剂化电子与 GO 上的部分离域 π 共轭，以稳定碳自由基，形成 π 键，使其在重新排列成最低能态时修复 GO 上的缺陷。

(9) 氨基酸还原

主要是 L-半胱氨酸还原。L-半胱氨酸的巯基易氧化形成二硫化物衍生物 L-胱氨酸，具有较强的还原性。Chen 等[94] 人用 L-半胱氨酸还原氧化石墨烯，GO 上的环氧基和羟基经两步双分子亲核反应，然后进行热消除。首先巯基释放质子提供亲核巯基，游离质子对羟基上的氧原子具有高亲和力，形成水分子离去，来还原 GO，同时 L-半胱氨酸被氧化为 L-胱氨酸。

L-谷胱甘肽（Glutathione，简称 GSH）是细胞中的天然抗氧化剂，能够减少许多活性氧。Pham 等[95] 人利用 GSH 的这一特性还原 GO，得到的石墨烯在含水和极性非质子溶剂如 THF、DMF 和 DMSO 中显示出良好的分散稳定性。GSH 封端起到了稳定和阻聚的作用，因为 GSH 能与 GO 表面上残余含氧官能团之间形成氢键。该研究认为，GSH 能够释放质子并且与另一种 GSH 发生分子间反应以形成谷胱甘肽二硫化物，释放的质子会与 GO 表面上存在的氧基团结合，随后作为水分子释放。

2.2.6 植物提取物还原

除了使用化学还原剂外，生物材料作为 GO 还原剂的应用近年来受到了广泛关注，与常规化学试剂相比，这种天然存在的还原剂更加环保。

Wang 等[96] 人报道了用绿茶还原氧化石墨烯，因为茶中含有的多酚以及多酚化合物邻苯三酚具有优异的抗氧化特性，其机理是酚类基团被氧化为相应的醌。事实上，多酚化合物具有合成和稳定 Au、Ag、Pd 和 Fe 微粒的能力。绿茶提取物还原得到石墨烯的电导率为 53S/m，并且在极性溶剂如乙醇、甲醇、丙酮、DMF、NMP、DMAC 和 DMSO 中具有良好的分散稳定性。

Tabrizi 等[97] 人报道了以玫瑰花作为还原剂，由于玫瑰水含有天然抗氧化剂，例如，酚类化合物和黄烷醇糖苷，所以玫瑰花瓣被用作还原剂，所得的石墨烯可以在水溶液中稳定保存一个月。

2.2.7 微生物还原

在环境中发现的一种厌氧菌（希瓦氏菌）有能力利用其呼吸过程中的无机或有机化合物作为末端电子受体，这些有机体也能够利用固体作为终端电子受体，此微生物可用来还原 GO。

Salas 等[98] 人用希瓦氏菌在严格无氧环境中还原 GO 得到还原氧化石墨烯。同时，Wang 等[99] 证明了在有氧条件下希瓦氏菌对 GO 也具有还原能力。

Akhavan 等[100] 人报道了大肠杆菌在厌氧条件下利用混合酸发酵途径来还原氧化石墨烯。该研究将 GO 样品和大肠杆菌培养 48 小时得到石墨烯，其薄膜面电阻为 $3.4 \times 10^9 \Omega/\square$（相比而言，GO：$2.9 \times 10^{11} \Omega/\square$）。他们认为，糖酵解过程中代谢产生的电子和氢参与了 GO 的还原。此外，研究发现 GO 为大肠杆菌的吸附和增殖提供了生物相容性位点，而还原后的 GO 则抑制了大肠杆菌的进一步增殖。

2.2.8 蛋白质还原

蛋白质中存在氨基等基团，对 GO 有一定的还原作用。Liu 等[101]报道了用牛血清白蛋白（Albumin Bovine Serum，简称 BSA）还原石墨烯氧化物，如图 2.8 所示。BSA 中酪氨酸的酚基团在高 pH 环境下电离并将电子转移至 GO。除了在碱性环境下对 GO 具有一定的脱氧能力，BSA 还能使复合物在较高离子氛下分散稳定存在。同时将 BSA-石墨烯氧化物与 Au、Pt、Pd、Ag 和乳胶纳米颗粒复合，可使纳米粒子与还原氧化石墨烯形成良好控制和良好分散的组装体。

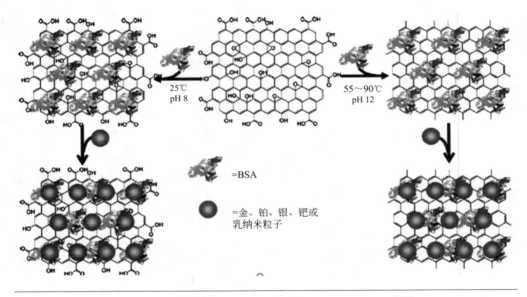

图 2.8　BSA 与 GO 稳定金属纳米离子作用机理示意[101]

2.2.9 激素还原

褪黑激素作为一种有效的自杀性抗氧化剂可以清除活性氧或氮。Esfandiar 等[102]用褪黑激素（N-乙酰基-5-甲氧基色胺）在 80℃氨（pH 9～10）条件下还原 GO 水悬浮液 3 小时后，得到石墨烯。反应混合物的高碱性阻止了石墨烯片的聚集并且增加了还原速率。此外，氧化的褪黑激素通过 π-π 相互作用吸附在石墨烯片表面上，有助于石墨烯的分散稳定性。研究认为，GO 表面上的环氧基团以两种不同的方式受褪黑激素亲核进攻：a. 首先形成共价 C-C 键，随后另一个褪黑激素分子转移氢通过脱水使羟基还原，所得的二聚体随后吸附在石墨烯片的表面上；b. 氮对酰氨基团的亲核攻击导致形成 C-N 共价键。另一方面，GO 表面上的羟基也以两种不同的方式被消除：a. 电子和氢的转移导致形成吲哚基阳离子基团，而通过脱水除去羟基处理；b. 在碱性条件下热处理时产生羟基和石墨烯自由基，羟基随后被褪黑激素阳离子自由基清除产生羟基褪黑激素化合物，随后清除第二羟基自由基以形成脱羟基褪黑激素化合物。

2.3 小结

将 GO 进行还原是大规模生产石墨烯的常用方法，也是利用石墨烯特殊的物理、化学、机械、热学和光学性能，并实现各种领域应用的重要路径。尽管利用氧化还原得到结构完美的石墨烯还难以实现，部分还原 GO 的方法已经得到了充分研究。本章介绍了氧化石墨的一些还原技术，并对一些方法的机理做了简单介绍。

上述对于 GO 还原的研究中积累了大量的实验现象、理论模拟结果，为 GO、还原氧化石墨、石墨烯的结构与性质提供了更清晰的见解，这有助于促进石墨烯材料的实际应用。由于各应用领域对石墨烯材料结构与性质具有不同的需求，未来对 GO 还原的研究应主要围绕关注两个主题：a. 更深入地了解还原机理；b. 如何控制石墨的氧化及相应 GO 的还原过程和产物形态。

─── 参考文献 ───

[1] Zhao Jinping, Pei Songfeng, Ren Wencai, et al. Efficient preparation of large-area graphene oxide sheets for transparent conductive films [J]. Acs Nano, **2010**, 4 (9): 5245-5252.

[2] Stankovich Sasha, Dikin Dmitriy A., Piner Richard D., et al. Synthesis of graphene-based nanosheets via chemical reduction of exfoliated graphite oxide [J]. Carbon, **2007**, 45 (7): 1558-1565.

[3] Gao Libo, Ren Wencai, Li Feng, et al. Total color difference for rapid and accurate identification of graphene [J]. Acs Nano, **2008**, 2 (8): 1625.

[4] Bae Sukang, Kim Hyeongkeun, Lee Youngbin, et al. Roll-to-roll production of 30-inch graphene films for transparent electrodes [J]. Nature Nanotechnology, **2010**, 5 (8): 574-578.

[5] Edwards Peter, Porch A, Jones Martin, et al. Basic materials physics of transparent conducting oxides [J]. Dalton Transactions, **2004**, 19 (19): 2995-3002.

[6] Lopez Vicente, Sundaram Ravi S, Gomez-Navarro Cristina, et al. Chemical Vapor Deposition Repair of Graphene Oxide: A Route to Highly Conductive Graphene Monolayers [J]. Advanced Materials, **2010**, 21 (46): 4683-4686.

[7] Novoselov K. S., Geim A. K., Morozov S, et al. Electric field effect in atomically thin carbon films [J]. Science, **2004**, 306 (5696): 666-669.

[8] Su Qi, Pang Shuping, Alijani Vajiheh, et al. Composites of Graphene with Large Aromatic Molecules [J]. Advanced Materials, **2010**, 21 (31): 3191-3195.

[9] Jeong Hae, Pyo Lee Yun, Lahaye R, et al. Evidence of Graphitic AB Stacking Order of Graphite Oxides

[J]. Journal of the American Chemical Society, **2008**, 130 (4): 1362-1366.

[10] Huang Xiao, Qi Xiaoying, Boey Freddy, et al. Graphene-based composites [J]. Chemical Society Reviews, **2012**, 41 (2): 666-686.

[11] Schniepp Hannes C, Je-Luen Li, Mcallister Michael J, et al. Functionalized single graphene sheets derived from splitting graphite oxide [J]. Journal of Physical Chemistry B, **2006**, 110 (17): 8535-8539.

[12] Mattevi Cecilia, Eda Goki, Agnoli Stefano, et al. Evolution of Electrical, Chemical, and Structural Properties of Transparent and Conducting Chemically Derived Graphene Thin Films [J]. Advanced Functional Materials, **2010**, 19 (16): 2577-2583.

[13] Robertson J and O'Reilly E. P. Electronic and atomic structure of amorphous carbon [J]. Phys Rev B Condens Matter, **1987**, 35 (6): 2946-2957.

[14] Beamson G. and Briggs D. High resolution XPS of organic polymers: the Scienta ESCA300 database [M]. City: Wiley, **1992**.

[15] Wu Zhong Shuai, Ren Wencai, Gao Libo, et al. Synthesis of high-quality graphene with a pre-determined number of layers [J]. Carbon, **2009**, 47 (2): 493-499.

[16] McAllister M. J., Li J. L., Adamson Douglas, et al. Single sheet functionalized graphene by oxidation and thermal expansion of graphite [J]. Chemistry of Materials, **2007**, 19 (18): 4396-4404.

[17] Wu Zhong-Shuai, Ren Wencai, Gao Libo, et al. Synthesis of graphene sheets with high electrical conductivity and good thermal stability by hydrogen arc discharge exfoliation [J]. Acs Nano, **2009**, 3 (2): 411-417.

[18] Kudin Konstantin N, Ozbas Bulent, Schniepp Hannes, et al. Raman Spectra of Graphite Oxide and Functionalized Graphene Sheets [J]. Nano Letters, **2007**, 8 (1): 36-41.

[19] A Becerril Héctor, Mao Jie, Liu Zunfeng, et al. Evaluation of solution-processed reduced graphene oxide films as transparent conductors [J]. Acs Nano, **2008**, 2 (3): 463-470.

[20] Yang Dongxing, Velamakanni Aruna, Bozoklu Gülay, et al. Chemical analysis of graphene oxide films after heat and chemical treatments by X-ray photoelectron and Micro-Raman spectroscopy [J]. Carbon, **2009**, 47 (1): 145-152.

[21] Pan Dengyu, Zhang Jingchun, Li Zhen, et al. Hydrothermal route for cutting graphene sheets into blue-luminescent graphene quantum dots [J]. Advanced Materials, **2010**, 22 (6): 734-738.

[22] Li Xiaolin, Wang Hailiang, T Robinson Joshua, et al. Simultaneous nitrogen doping and reduction of graphene oxide [J]. Journal of the American Chemical Society, **2009**, 131 (43): 15939-15944.

[23] Wang Xuan, Zhi Linjie, Müllen Klaus. Transparent, Conductive Graphene Electrodes for Dye-Sensitized Solar Cells [J]. Nano Letters, **2008**, 8 (1): 323-327.

[24] Zhu Yanwu, Murali Shanthi, Stoller Meryl D., et al. Microwave assisted exfoliation and reduction of graphite oxide for ultracapacitors [J]. Carbon, **2010**, 48 (7): 2118-2122.

[25] Hassan Hassan, Abdelsayed Victor, Khder Abd el Rahman, et al. Microwave synthesis of graphene sheets supporting metal nanocrystals in aqueous and organic media [J]. Journal of Materials Chemistry, **2009**, 19 (23): 3832-3837.

[26] Cote Laura J, Rodolfo Cruz Silva, Jiaxing Huang. Flash reduction and patterning of graphite oxide and its polymer composite [J]. Journal of the American Chemical Society, **2009**, 131 (31): 11027-11032.

[27] Zhang Yonglai, Guo Li, Wei Shu, et al. Direct imprinting of microcircuits on graphene oxides film by femtosecond laser reduction [J]. Nano Today, **2010**, 5 (1): 15-20.

[28] Voiry Damien, Yang Jieun, Kupferberg Jacob, et al. High-quality graphene via microwave reduction of solution-exfoliated graphene oxide [J]. Science, **2016**, 353 (6306): 1413-1416.

[29] Graeme Williams, Brian Seger, Kamat Prashant V. TiO$_2$-graphene nanocomposites. UV-assisted photocatalytic reduction of graphene oxide [J]. Acs Nano, **2008**, 2 (7): 1487.

[30] Kamat Prashant V. Photochemistry on Nonreactive and Reactive (Semiconductor) Surfaces [J]. Chemical Reviews, **1993**, 93 (1): 267-300.

[31] Kamat Prashant V., Bedja Idris, Hotchandani Surat. Photoinduced Charge Transfer between Carbon and Semiconductor Clusters. One-Electron Reduction of C60 in Colloidal TiO$_2$ Semiconductor Suspensions [J]. Journal of Physical Chemistry B, **1994**, 98 (37): 9137-9142.

[32] Anusorn Kongkanand, Kamat Prashant V. Electron storage in single wall carbon nanotubes. Fermi level equilibration in semiconductor-SWCNT suspensions [J] Acs Nano, **2007**, 1 (1): 13.

[33] Zhang Hao, Lv Xiaojun, Li Yueming, et al. P25-graphene composite as a high performance photocatalyst [J]. Acs Nano, **2010**, 4 (1): 380-386.

[34] Kim Sung Ryong, Parvez Md. Khaled, Chhowalla Manish. UV-reduction of graphene oxide and its application as an interfacial layer to reduce the back-transport reactions in dye-sensitized solar cells [J]. Chemical Physics Letters, **2009**, 483 (1): 124-127.

[35] Yang Nailiang, Zhai Jin, Wang Dan. Two-Dimensional Graphene Bridges Enhanced Photoinduced Charge Transport in Dye-Sensitized Solar Cells [J]. ACS Nano, **2010**, 4 (2): 887-894.

[36] Graeme Williams, Kamat Prashant V. Graphene-semiconductor nanocomposites: excited-state interactions between ZnO nanoparticles and graphene oxide [J]. Langmuir, **2009**, 25 (24): 13869-13873.

[37] Yun Hau Ng, Iwase Akihide, Kudo Akihiko, et al. Reducing Graphene Oxide on a Visible-Light BiVO$_4$ Photocatalyst for an Enhanced Photoelectrochemical Water Splitting [J]. Journal of Physical Chemistry Letters, **2010**, 1 (17): 2607-2612.

[38] Zhou Yong, Bao Qiaoliang, Tang Lena Ai Ling, et al. Hydrothermal Dehydration for the "Green" Reduction of Exfoliated Graphene Oxide to Graphene and Demonstration of Tunable Optical Limiting Properties [J]. Chemistry of Materials, **2009**, 21 (13): 2950-2956.

[39] Wang Hailiang, Tucker Robinson Joshua, Li Xiaolin, et al. Solvothermal reduction of chemically exfoliated graphene sheets [J]. Journal of the American Chemical Society, **2009**, 131 (29): 9910.

[40] Sergey Dubin, Scott Gilje, Kan Wang, et al. A one-step, solvothermal reduction method for producing reduced graphene oxide dispersions in organic solvents [J]. Acs Nano, **2010**, 4 (7): 3845-3852.

[41] Demazeau Gérard. Solvothermal processes: A route to the stabilization of new material [J]. Journal of Materials Chemistry, **1999**, 9 (1): 15-18.

[42] Sun Xiaoming, Li Yadong. Colloidal carbon spheres and their core/shell structures with noble-metal nanoparticles [J]. Angew Chem Int Ed Engl, **2004**, 43 (5): 597-601.

[43] Zhang Li Wu, Fu Hong Bo, Zhu Yong Fa. Efficient TiO$_2$ Photocatalysts from Surface Hybridization of TiO$_2$ Particles with Graphite-like Carbon [J]. Advanced Functional Materials, **2010**, 18 (15): 2180-2189.

[44] Luo Lin-Bao, Yu Shu-Hong, Qian Hai-sheng, et al. Large-Scale Fabrication of Flexible Silver/Cross-Linked PoIy (vinyl alcohol) Coaxial Nanocables by a Facile Solution Approach [J]. Journal of the American Chemical Society, **2005**, 127 (9): 2822-2823.

[45] Zhou Ming, Wang Yuling, Zhai Yueming, et al. Controlled synthesis of large-area and patterned electrochemically reduced graphene oxide films [J]. Journal of Molecular Structure, **2010**, 15 (25): 6116-6120.

[46] Zhang Xi Zhao, Wang Guang Yi. Direct Electrochemical Reduction of Single-Layer Graphene Oxide and Subsequent Functionalization with Glucose Oxidase [J]. Journal of Physical Chemistry C, **2009**, 113

(32): 14071-14075.

[47] Ramesha Ganganahalli K, Sampath Srinivasan. Electrochemical Reduction of Oriented Graphene Oxide Films: An in Situ Raman Spectroelectrochemical Study [J]. J. phys. chem. c, **2009**, 113 (19): 7985-7989.

[48] Muszynski Ryan, Seger Brian, Kamat Prashant V. Decorating graphene sheets with gold nanoparticles [J]. J. phys. chem. c, **2008**, 112 (14): 5263-5266.

[49] Si Yongchao, Samulski Ed. Synthesis of water soluble graphene [J]. Nano Letters, **2008**, 8 (6): 1679.

[50] Gao Wei, Alemany Lawrence B., Ci Lijie, et al. New insights into the structure and reduction of graphite oxide [J]. Nature Chemistry, **2009**, 1 (5): 403-408.

[51] Shin Hyeon Jin, Kim Ki Kang, Benayad Anass, et al. Efficient Reduction of Graphite Oxide by Sodium Borohydride and Its Effect on Electrical Conductance [J]. Advanced Functional Materials, **2010**, 19 (12): 1987-1992.

[52] Chua Chun Kiang, Pumera Martin. Reduction of graphene oxide with substituted borohydrides [J]. Journal of Materials Chemistry A, **2013**, 1 (5): 1892-1898.

[53] Pham Viet Hung, Hur Seung, Kim Eui, et al. Highly efficient reduction of graphene oxide using ammonia borane [J]. Chemical Communications, **2013**, 49 (59): 6665-6667.

[54] Ambrosi Adriano, Chua Chun Kiang, Bonanni Alessandra, et al. Lithium Aluminum Hydride as Reducing Agent for Chemically Reduced Graphene Oxides [J]. Chemistry of Materials, **2012**, 24 (12): 2292-2298.

[55] Moon In, Junghyun Lee, Ruoff Rodney, et al. Reduced graphene oxide by chemical graphitization [J]. Nature Communications, **2010**, 1 (6): 73.

[56] Cui Peng, Junghyun Lee, Hwang Eunhee, et al. One-pot reduction of graphene oxide at subzero temperatures [J]. Chemical Communications, **2011**, 47 (45): 12370-12372.

[57] Pei Songfeng, Zhao Jinping, Du Jinhong, et al. Direct reduction of graphene oxide films into highly conductive and flexible graphene films by hydrohalic acids [J]. Carbon, **2010**, 48 (15): 4466-4474.

[58] Chen Yao, Zhang Xiong, Zhang Dacheng, et al. High performance supercapacitors based on reduced graphene oxide in aqueous and ionic liquid electrolytes [J]. Carbon, **2011**, 49 (2): 573-580.

[59] Chua Chun Kiang, Pumera Martin Renewal of sp^2 bonds in graphene oxides via dehydrobromination [J]. Journal of Materials Chemistry, **2012**, 22 (43): 23227-23231.

[60] Chua Chun Kiang, Ambrosi Adriano, Pumera Martin. Graphene oxide reduction by standard industrial reducing agent: Thiourea dioxide [J]. Journal of Materials Chemistry, **2012**, 22 (22): 11054-11061.

[61] Wang Yanqing, Sun Ling, Fugetsu Bunshi. Thiourea Dioxide as a Green Reductant for the Mass Production of Solution-Based Graphene [J]. Bulletin of the Chemical Society of Japan, **2012**, 85 (12): 1339-1344.

[62] Ma Qi, Song Jinping, Jin Chun, et al. A rapid and easy approach for the reduction of graphene oxide by formamidinesulfinic acid [J]. Carbon, **2013**, 54 (8): 36-41.

[63] Node M., et al. ChemInform Abstract: A convenient reduction of functionalized polycyclic aromatics into parent hydrocarbons [J]. Chemischer Informationsdienst, **1982**, 13 (25): 689-692.

[64] Chun Kiang Chua, Martin Pumera. Selective removal of hydroxyl groups from graphene oxide [J]. Chemistry of Materials, **2013**, 19 (6): 2005-2011.

[65] Kollenz G, Penn G, Theuer R, et al. Reactions of cyclic oxalyl compounds—38. New isoindigoide dyes from heterocyclic 2, 3-diones—Synthesis and thermal rearrangement [J]. Tetrahedron, **1996**, 52 (15): 5427-5440.

[66] Liu Hongtao, Zhang Lei, Guo Yunlong, et al. Reduction of graphene oxide to highly conductive graphene by Lawesson's reagent and its electrical applications [J]. Journal of Materials Chemistry C, **2013**, 1 (18): 3104-3109.

[67] Zhou Tiannan, Chen Feng, Liu Kai, et al. A simple and efficient method to prepare graphene by reduction of graphite oxide with sodium hydrosulfite [J]. Nanotechnology, **2011**, 22 (4): 045704.

[68] Liu Yanzhen, Li Yongfeng, Yang Yonggang, et al. Reduction of graphene oxide by thiourea [J]. J Nanosci Nanotechnol, **2011**, 11 (11): 10082-10086.

[69] Some Surajit, Kim Youngmin, Yoon Yeoheung, et al. High-quality reduced graphene oxide by a dual-function chemical reduction and healing process [J]. Sci Rep, **2013**, 3 (1929): 01929.

[70] Chan Kim Min, Hwang Gyeong S, Ruoff Rodney S. Epoxide reduction with hydrazine on graphene: a first principles study [J]. Journal of Chemical Physics, **2009**, 131 (6): 064704.

[71] Park Sungjin, An Jinho, Jung Inhwa, et al. Colloidal suspensions of highly reduced graphene oxide in a wide variety of organic solvents [J]. Nano Letters, **2009**, 9 (4): 1593-1597.

[72] Viet Hung Pham, Cuong Tran, Nguyen-Phan Thuy-Duong, et al. One-step synthesis of superior dispersion of chemically converted graphene in organic solvents [J]. Chemical Communications, **2010**, 46 (24): 4375-4377.

[73] Zhou Xuejiao, Zhang Jiali, Wu Haixia, et al. Reducing Graphene Oxide via Hydroxylamine: A Simple and Efficient Route to Graphene [J]. Journal of Physical Chemistry C, **2011**, 115 (24): 11957-11961.

[74] Mao Shun, Yu Kehan, Cui Shumao, et al. A new reducing agent to prepare single-layer, high-quality reduced graphene oxide for device applications [J]. Nanoscale, **2011**, 3 (7): 2849-2853.

[75] Sakura Takeshi, Nagasaki Yukio Preparation of gold colloid using pyrrole-2-carboxylic acid and characterization of its particle growth [J]. Colloid Polymer Science, **2007**, 285 (12): 1407-1410.

[76] Amarnath Chellachamy Anbalagan, Hong Chang Eui, Kim Nam Hoon, et al. Efficient synthesis of graphene sheets using pyrrole as a reducing agent [J]. Carbon, **2011**, 49 (11): 3497-3502.

[77] Liu Sen, Tian Jingqi, Wang Lei, et al. A method for the production of reduced graphene oxide using benzylamine as a reducing and stabilizing agent and its subsequent decoration with Ag nanoparticles for enzymeless hydrogen peroxide detection [J]. Carbon, **2011**, 49 (10): 3158-3164.

[78] Chen Yao, Zhang Xiong, yu Peng, et al. Stable dispersions of graphene and highly conducting graphene films: a new approach to creating colloids of graphene monolayers [J]. Chemical Communications, **2009**, 30 (30): 4527-4529.

[79] Che Jianfei, Shen Liying, Xiao Yinghong. A new approach to fabricate graphene nanosheets in organic medium: Combination of reduction and dispersion [J]. Journal of Materials Chemistry, **2010**, 20 (9): 1722-1727.

[80] Lei Zhibin, Lu Li, Zhao X. S.. The Electrocapacitive Properties of Graphene Oxide Reduced by Urea [J]. Energy Environmental Science, **2012**, 5 (4): 6391-6399.

[81] Su Peng, Guo Huilin, Tian Lei, et al. An efficient method of producing stable graphene suspensions with less toxicity using dimethyl ketoxime [J]. Carbon, **2012**, 50 (15): 5351-5358.

[82] Dreyer Daniel R., Murali Shanthi, Zhu Yanwu, et al. Reduction of graphite oxide using alcohols [J]. Journal of materials chemistry, **2011**, 21 (10): 3443-1447.

[83] Gao Jian, Liu Fang, Liu Yiliu, et al. Environment-Friendly Method To Produce Graphene That Employs Vitamin C and Amino Acid [J]. Chemistry of Materials, **2010**, 22 (7): 2213-2218.

[84] Zhang Jiali, Yang Haijun, Shen Guangxia, et al. Reduction of graphene oxide via L-ascorbic acid [J]. Chemical Communications, **2010**, 46 (7): 1112.

[85] Zhu Chengzhou, Guo Shaojun, Fang Youxing, et al. Reducing sugar: new functional molecules for the green synthesis of graphene nanosheets [J]. Acs Nano, **2010**, 4 (4): 2429-2437.

[86] Kim Young-Kwan, Kim Mi-Hee, Min Dal-Hee. Biocompatible reduced graphene oxide prepared by using dextran as a multifunctional reducing agent [J]. Chemical Communications, **2011**, 47 (11): 3195-3197.

[87] Fan Zhuangjun, Wang Kai, Wei Tong, et al. An environmentally friendly and efficient route for the reduction of graphene oxide by aluminum powder [J]. Carbon, **2010**, 48 (5): 1686-1689.

[88] Fan Zhuang-Jun, Kai Wang, Yan Jun, et al. Facile synthesis of graphene nanosheets via Fe reduction of exfoliated graphite oxide [J]. Acs Nano, **2011**, 5 (1): 191-198.

[89] Mei Xiaoguang, Ouyang Jianyong. Ultrasonication-assisted ultrafast reduction of graphene oxide by zinc powder at room temperature [J]. Carbon, **2011**, 49 (15): 5389-5397.

[90] Ramendra Sundar Dey, Saumen Hajra, Sahu Ranjan K, et al. A rapid room temperature chemical route for the synthesis of graphene: metal-mediated reduction of graphene oxide [J]. Chemical Communications, **2012**, 48 (12): 1787-1789.

[91] Barman Barun Kumar, Mahanandia Pitamber, and Nanda Karuna Kar Instantaneous reduction of graphene oxide at room temperature [J]. Rsc Advances, **2013**, 3 (31): 12621-12624.

[92] Liu Yanzhen, Li Yongfeng, Zhong Ming, et al. A green and ultrafast approach to the synthesis of scalable graphene nanosheets with Zn powder for electrochemical energy storage [J]. Journal of Materials Chemistry, **2011**, 21 (39): 15449-15455.

[93] Feng Hongbin, Cheng Rui, Zhao Xin, et al. Corrigendum: A low-temperature method to produce highly reduced graphene oxide [J]. Nature Communications, **2013**, 4 (2): 1539.

[94] Chen Dezhi, Li Lidong, Guo Lin. An environment-friendly preparation of reduced graphene oxide nanosheets via amino acid [J]. Nanotechnology, **2011**, 22 (32): 325601.

[95] Pham Tuan Anh, Kim Jongsik, Kim Jong Su, et al. Corrigendum to "One-step reduction of graphene oxide with l-glutathione" [Colloids Surf. A: Physicochem. Eng. Aspects 384 (2011) 543 – 548] [J]. Colloids Surfaces A: Physicochemical Engineering Aspects, **2011**, 1 (386): 195.

[96] Wang Yan, Shi Zixing, Yin Jie. Facile Synthesis of Soluble Graphene via a Green Reduction of Graphene Oxide in Tea Solution and Its Biocomposites [J]. Acs Appl Mater Interfaces, **2011**, 3 (4): 1127-1133.

[97] Tabrizi Mahmoud Amouzadeh. Green-synthesis of reduced graphene oxide nanosheets using rose water and a survey on their characteristics and applications [J]. Rsc Advances, **2013**, 3 (32): 13365-13371.

[98] Salas Everett, Sun Zhengzong, Luttge Andreas, et al. Reduction of graphene oxide via bacterial respiration [J]. Acs Nano, **2010**, 4 (8): 4852-4856.

[99] Wang Gongming, Qian Fang, Saltikov Chad, et al. Microbial Reduction of Graphene Oxide by Shewanella [J]. Nano Research, **2011**, 4 (6): 563-570.

[100] Akhavan O., Ghaderi E. Escherichia coli bacteria reduce graphene oxide to bactericidal graphene in a self-limiting manner [J]. Carbon, **2012**, 50 (5): 1853-1860.

[101] Liu Jinbin, Fu Songhe, Yuan Bin, et al. Toward a universal "adhesive nanosheet" for the assembly of multiple nanoparticles based on a protein-induced reduction/decoration of graphene oxide [J]. Journal of the American Chemical Society, **2010**, 132 (21): 7279.

[102] Esfandiar Ali., Akhavan O., Irajizad A.. Melatonin as a powerful bio-antioxidant for reduction of graphene oxide [J]. Journal of Materials Chemistry, **2011**, 21 (29): 10907-10914.

氧化石墨烯的修饰和改性

第3章
CHAPTER 3

经 Hummers' 法等氧化方法以及不同的还原方法，可以规模化制备 GO 和石墨烯。多年研究认为，GO 结构有多种可能构型，其中得到广泛认同的一种是"GO 是具有蜂窝状主体的二维片层结构，表面和边缘存在着大量的含氧官能团"。表征发现，在 GO 表面上主要有环氧基和羟基，而在边缘上主要有羟基、羧基、羰基、酯基（五元环、七元环）等缺陷结构存在。尽管这些缺陷在不同程度上破坏了石墨烯的共轭结构，改变了石墨烯结构的若干本征特性，但官能团的存在却为 GO 功能化提供了化学反应基础，活性官能团作为化学反应活性点为 GO 的修饰与改性提供了重要的科学支撑。

随着材料领域的发展，人们对于材料的功能性需求更为严苛，迫切需要在交通运输、建筑材料、能量存储与转化等领域应用性质更加优良的材料出现，石墨烯以优异的声、光、热、电、力等性质引起关注，作为前驱体的 GO 也以其灵活的物理化学性质、可规模化制备的特点更成为应用基础研究的热点。虽然 GO 具有诸多独特性质，但是由于范德华作用以及 π-π 作用等强相互作用力，使 GO 之间很容易在不同体系中发生团聚，导致很多在纳米尺度上表现的优异性能随着 GO 片层的聚集显著降低直至消失，阻碍了 GO 的进一步应用。因而，解决 GO 在不同介质中的解离和分散等问题是实现 GO 广泛应用的重要前提。此外，不同的应用体系往往要不同的功能体现和界面结合等特征，故而要经常对 GO 表面进行修饰改性。GO 本身含有丰富的含氧官能团，也可在 GO 表面引入其他功能基团，或者利用 GO 之间和 GO 与其他物质间的共价键或非共价键作用进行化学反应接枝其他官能团。由于

GO 结构的不确定性，导致采用化学方式对 GO 进行修饰与改性机理复杂化，很难得到结构单一的产品。

尽管存在上述困难，对 GO 进行修饰改性仍然是 GO 相关材料应用能否实现稳定、可控规模化应用的关键，因此对常用方法和技术进行总结非常必要。根据现有的研究成果，目前对于 GO 的修饰与改性主要包含以下方式和方法。

3.1 氧化石墨烯的还原改性

前一章已经总结了关于 GO 还原的方法和技术。GO 经还原处理后，对于提高其导电性、比表面积等大有裨益，使得石墨烯可以应用于对导电性、导热性等要求更高的应用中。

在 GO 还原过程中，含氧官能团的去除和控制过程本身也可成为石墨烯改性的一种方式，通过不同还原方式得到的还原氧化石墨烯，其结构、形貌、组分可通过还原条件进行适当的调控。比如，Song 等[1]人介绍了氩气氛围下，将 GO 在 1100~2000℃的温度范围内进行热处理，并对得到的石墨烯结构和吸附性能进行研究，发现制备的石墨烯粉体材料的比表面积增加至前驱体材料的四倍以上，并提高了 GO 的导热性能，增强了 GO 材料热管理方面的性能。此外，采用微波处理可实现对 GO 及其衍生物修复缺陷。Longo 等[2]人报道了微波加热法制备 GO 胶体分散体在 N-甲基-2-吡咯烷酮中的还原过程。利用光催化还原 GO、制备 GO 复合材料可以成为一种绿色环保的方法。2008 年，Williams 等[3]报道了 GO 与纳米 TiO_2 粒子悬浮在乙醇中经紫外光辐照还原可得到 TiO_2 复合石墨烯薄膜，相比较 TiO_2 粒子，该固体薄膜横向电阻降低一个数量级。另外，利用化学试剂进行还原后，得到的石墨烯往往含有还原剂的某些成分。这样通过不同的还原方式，GO 可能得到相应的修饰或者改性，使其用途和特性更加广泛，很多还原过程可以直接提升石墨烯在特定溶剂中的分散性或者在特定复合体系中的其他功能特性。

除了本部分外，以下主要总结针对 GO 结构本身的特定目标修饰改性技术。

3.2 共价修饰改性

共价键修饰主要是基于 GO 表面丰富的含氧官能团与功能分子之间的化学反应[4]，该功能分子可以是小分子，也可以是大分子聚合物。采用共价键修饰 GO 所得到的复合材料可以提高 GO 与其他高分子材料或其他有机小分子的相容性，从而得到以 GO 主导或者以其他物质主导的稳定复合材料。共价键修饰可分别从羧基、环氧基、羟基三个角度设计化学反应，得到目标物修饰的 GO。

3.2.1 羧基修饰

一般认为羧基分布在 GO 片层的边缘，其化学反应活性最高，羧基能与二氯亚砜（$SOCl_2$）酰基化反应[5]、碳化二亚胺[6]、N,N'-二环己基碳二亚胺[7]氨基化以及醇类发生酯化[8] 等反应，其中 $SOCl_2$ 和含有两个及以上的氨基小分子反应属于对 GO 的活化处理。活化后的 GO 再与不同的亲核试剂发生偶联反应，通过该类型的亲核反应得到共价键，在 GO 表面添加了其他功能性的基团，从而赋予 GO 复合材料相应的功能，增强其在光学[9]、催化[10]、生物器件[11]、药物靶向运输[12]、超级电容器[13]、复合高分子材料[14,15] 等应用中的性能。例如，将羧基酯化的主要目的是将脂肪族的碳链或小分子修饰在 GO 表面，从而增强其在有机溶剂中的溶解性[16]。更复杂的羧基修饰反应，先对羧基与多氨基化合物修饰，然后小分子之间发生交联反应，达到在 GO 表面修饰聚合物的效果[7]。此外，异氰酸衍生物与羧基反应活性也较高[8]，异氰酸衍生物与 GO 羧基反应生成 GO 表面修饰酰氨基酸酯类物质，在极性非质子性有机溶剂中的分散性较好。

3.2.2 环氧基团修饰

环氧基团是 GO 表面最丰富的含氧官能团之一。环氧基团不稳定，易发生开环反应，在酸性条件下开环形成羟基，碱性条件下易发生亲核取代反应。例如，利用 GO 上的环氧基团与十八胺开环反应形成官能化的 GO 复合物，其热稳定性比普通环氧树脂高，同时机械性能明显增强[17]。在制作导电石墨烯薄膜时[18]，利用正己胺进攻 GO 表面的环氧基团，发生开环反应生成氨基化的 GO，然后经还原形成导电石墨烯纸。Jiang 等人[19] 用环氧与氨基开环反应在 GO 表面修饰 β-环糊精，然后与嵌段共聚物通过非共价键构筑三维网络结构，得到的水凝胶在高温下显示快速溶胶凝胶转变，如图 3.1 所示。在碱性（KOH）条件下，GO 的环氧基团通过与离子液体 [1-(3-氨基丙基)-3-甲基咪唑溴胺，RNH_2] 发生亲核开环反应，使 GO 被修饰上氨基[20]，得到的 p-CCG 可在不添加任何表面活性剂或者大分子稳定剂条件下长时间稳定分散在水、DMF、DMSO 等溶剂中。以离子化合物修饰后的 GO 可以应用在防腐[21]、储能[22]、传感器[23] 等领域。

受合成聚合物化学或生物化学的启发，有报道研究 GO 和聚烯丙胺或硼酸钠通过环氧基和羟基交联[24,25]，所得到的 GO 膜机械性能明显增强。缺点是这些交联方式无法使 GO 在溶剂中稳定存在，所得到的 GO 膜在与水接触后会发生溶胀而破裂。

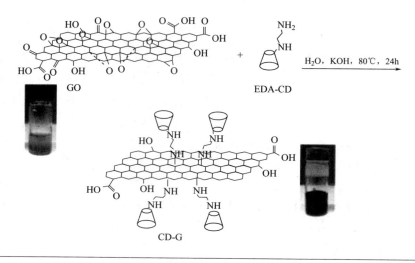

图 3.1 GO 以 β-环糊精修饰[19]

3.2.3 羟基修饰反应

GO 表面的羟基含量丰富，羟基作为活性官能团也能与氨基、酰基、异氰酸等发生化学反应，但由于羟基的反应活性次于羧基和环氧，在发生这些反应时主要以羧基和环氧为主体。利用羟基修饰主要是缩合反应类型修饰，例如利用 GO 的羟基与 3-氨基丙基三乙氧基硅烷的共价官能化，合成功能化的 GO（f-GO）[26]，然后通过浸渍沉淀技术诱导的相转化，加入不同比例的 GO 和 f-GO 制备出杂化聚偏二氟乙烯（Polyvinylidene Difluoride，简称 PVDF）超滤膜。同时，羟基也可以作为亲核试剂攻击酮，例如，2-溴-2-甲基丙酰溴与 GO 上的羟基反应可形成用于原子转移自由基聚合（ATRP）的引发剂，引发 MMA 单体在 GO 片上聚合生成 GO-聚甲基丙烯酸甲酯（GO-PMMA）[27]。

Mcgrail 等[28] 将羟基与腈在酸性水溶液中进行快速反应实现修饰改性，如图 3.2 所示。其优点包括：a. 可以在含有各种小分子及含腈基的小分子或聚合物的水溶液中容易且迅速地反应；b. 产物可以通过离心或过滤等简单的分离方式快速得到，并保持可调节的溶解性和功能性。

GO 共价修饰方法主要应用于制备复合材料。利用不同类型的聚合物修饰，将 GO 表面形成聚合物刷，而后经不同的方式组装得到一维、二维、三维复合物材料，得到多重功能或者性能增强的新材料。关于如何与聚合物形成复合材料的方法会在下一小结中详细介绍。

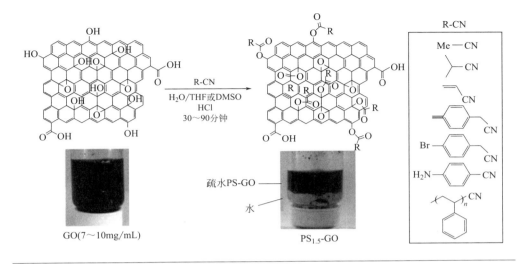

图 3.2 羟基与腈基反应修饰 GO[28]

3.3 聚合物共价修饰氧化石墨烯表面

利用 GO 表面的羟基、环氧基、羧基等官能团，有多种方法将其与高分子化合物进行化学键合。根据反应过程，这类修饰可分为两类：a. 从 GO 表面开始，原位聚合制备 GO-高分子材料（Graft-from GO）；b. 将制得的高分子链与 GO 链接（Graft-to GO）。

3.3.1 Graft from 接枝修饰方式

Graft from 方法利用 GO 表面的活性含氧官能团作为反应活性位点，或者将 GO 进行修饰后产生引发单体聚合的活性位点，进而引发单体在 GO 表面发生原位聚合反应。该方法主要优势在于：发生聚合反应的活性位点丰富，形成聚合物分子刷接枝在 GO 表面，易产生高接枝率的 GO 与聚合物的复合物。利用此种方法得到的聚合物接枝率较高，缺点在于聚合物的分子量以及分子量分布不容易控制。Fang 等[29]人报道了以 Graft from 的方式得到聚苯乙烯（Polystyrene，简称 PS）修饰的纳米石墨烯。实验中，以 GO 为前驱体用水合肼进行弱还原，得到羟基化的石墨烯纳米片，引发剂分子通过重氮加成与羟基化的石墨烯表面共价键合，达到了 82% 的 PS 接枝率。聚苯乙烯-石墨烯复合材料拉伸强度和杨氏模量分别比对比样品增加约 70% 和 57%。

此外，还可以用格氏试剂将 GO 表面的羟基、环氧基和羧基格氏化，然后与 $TiCl_4$ 反应制备 Ziegler-Natta 催化剂。利用改性过的催化剂，经原位催化丙烯在 GO 表面聚合生成聚丙烯（Polypropylene，简称 PP-g-GO）复合材料，如图 3.3 所示[30]。该复合材料在 PP 树脂中可实现均相分散，因为 PP 长链的存在避免了 GO 在 PP 中的团聚问题。PP-g-GO 在进一步的高温（190℃）加工过程中，GO 被初步还原，从而提高复合材料的导电性。通过这种原位聚合的方式，仅需 1.52%（质量）的 GO 添加量即可达到导静电的水平（电导率 10^{-6} S/m）。

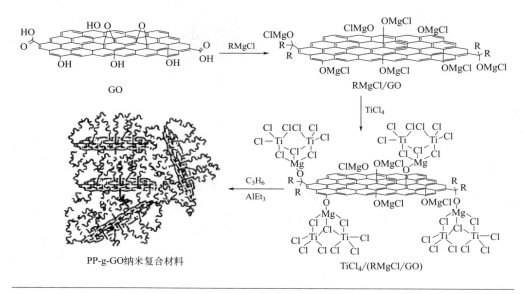

图 3.3 原位聚合 PP-g-GO 示意[30]

将 GO 上羧基改性成羟基，或直接利用 GO 上的羟基，与原子转移自由基聚合（ATRP）引发前体（2-溴-2-甲基丙酰溴）反应，可制得改性 GO 引发前体。利用该前体，可以引发大多数烯类单体（如丙烯酸类、苯乙烯类等）的自由基聚合，合成 GO 改性的高分子材料，如图 3.4 所示[31]。合成的复合材料常用作增强填料，例如 GO-g-PMMA 可在 PMMA 中均匀分散，大幅提高 PMMA 的韧性。只用 1%（质量）的 GO-g-PMMA 添加量，就可使纯 PMMA 的断裂伸长率提高 1 倍，杨氏模量提高 16%；而单纯 1%（质量）的 GO 则对 PMMA 断裂伸长率几乎没影响，杨氏模量反而降低 22%[27]。

Ma 等[32]人直接利用 GO 表面官能团-OH 产生自由基，引发烯类单体聚合，从而在 GO 表面接枝聚合物。其过程是以铈盐产生自由基，引发单体聚合，该方法的优势在于操作简单且接枝率随着单体浓度变化而变化，但由于存在链转移、双基终止和偶合终止等，导致聚合物的分子量无法精确控制。

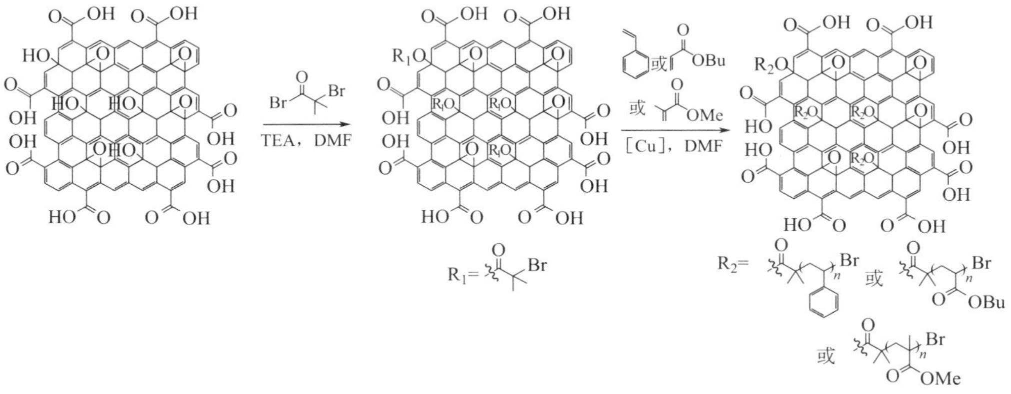

图 3.4 ATRP 合成 GO/烯类高分子复合材料[31]

3.3.2 Graft to 修饰改性氧化石墨烯表面

典型的 Craft to 的修饰方式采取"点击化学"（Click Chemistry）的方式，将 GO 炔基化或者叠氮化后与大分子发生点击反应。点击法可分为边缘点击和表面点击两种。边缘点击一般是通过化学反应活化羧基，或者通过催化剂使功能化反应物与羧基之间通过共价键连接生成酯或酰胺，从而对 GO 进行边缘点击功能化改性。而表面点击则是利用 GO 表面的环氧羟基等得到叠氮或者炔基化的 GO 后，再与聚合物点击反应。采用 Graft to 方式对 GO 进行接枝修饰，其主要优势在于所接枝聚合物分子量大小的可控性，同时还可将不同类型的目标聚合物接枝在 GO 表面，对 GO 应用的扩展十分有利。该方式的主要局限性在于聚合物分子与 GO 的空间位阻以及分子间作用力等因素影响，将直接导致聚合物的接枝率偏低。

以下介绍几个典型的边缘点击方式修饰接枝目标聚合物。

Sun 等[33]人使用点击边缘功能化法对 GO 修饰改性，在 GO 边缘引入炔基后与叠氮化合物发生点击反应。实验过程是：70℃下，$SOCl_2$ 与 GO 回流 24 小时，活化 GO 边缘上的羧基生成酰氯，而后，室温下与炔丙醇 24 小时进行酯化反应，得到炔基化的 GO（GO—C≡CH），再与叠氮基聚苯乙烯经点击反应后将含有叠氮基团聚苯乙烯接枝在 GO 上，从而增强 GO 在有机溶剂中的溶解分散性。

类似的，2010 年 Kou 等[34]采用边缘点击化方式将不同含有叠氮基团的聚合物引入 GO，通过酰胺化的方法得到叠氮基化的 $GO-N_3$。在 1-(3-二甲氨基丙基)-3-乙基碳二亚胺和 N-羟基琥珀酰亚胺（NHS）的催化作用下，GO 与 3-叠氮基丙胺

($NH_2CH_2CH_2CH_2N_3$) 在室温反应 24 小时,生成 GO-N_3。GO-N_3 与含炔基的聚乙二醇 (PEG—C≡CH)、聚苯乙烯 (PS—C≡CH)、甘氨酸 (Gly—C≡CH)、苯丙氨酸 (Phe—C≡CH) 以及棕榈酸炔丙酯等,发生点击反应。

Pan 等[35]人采用"点击化学"法将聚 (N-异丙基丙烯酰胺)(PNIPAM) 接枝到 GO 上,作为药物载体应用于药物传输领域。其条件是:在 N,N'-二环己基碳二亚胺和 4-二甲氨基吡啶 (DMAP) 的催化作用下,将 GO 与炔丙胺 (NH_2CH_2C≡CH) 在室温下搅拌 16 小时,得到 GO—C≡CH,GO—C≡CH 与叠氮化的 PNIPAM,通过点击化学反应在 GO 上接枝一个具有良好热响应性的聚合物分子 PNIPAM,得到温敏性复合材料 PNIPAM-GOS。如图 3.5 所示,在复合材

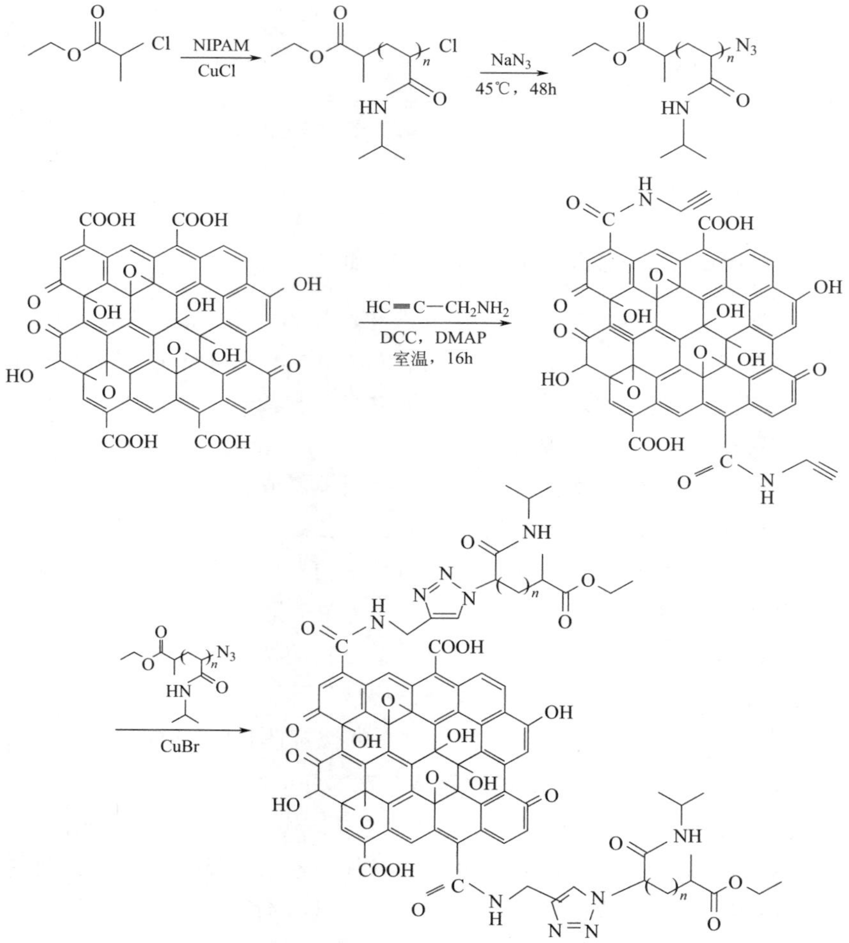

图 3.5 PNIPAM-GOS 的合成路线[35]

料中 PNIPAM 约占 50%，同时该复合物在生理盐溶液中具有良好的溶解性和稳定性。

关于"表面点击"的研究多有报道。Salvio 等[36] 报道了用"表面点击"的方式改性 GO，与之前所述一致，用叠氮基与炔基发生 Click 反应得到长烷基链修饰的 GO 复合物。作者将 GO 分散在水和乙腈比例为 1∶1 的混合液中，在 N_2 保护下与 NaN_3 加热回流 7 天，得到 GO 的叠氮衍生物（GO-N_3）与十八烷炔（CH≡C$C_{16}H_{33}$）在室温下搅拌 10 天，反应得到 GO 的烷基化改性复合材料（GO-C_{18}）。如图 3.6 所示，得到的 GO-C_{18} 含有长烷基链的复合物，增强了 GO 在非极性溶剂中的分散性。

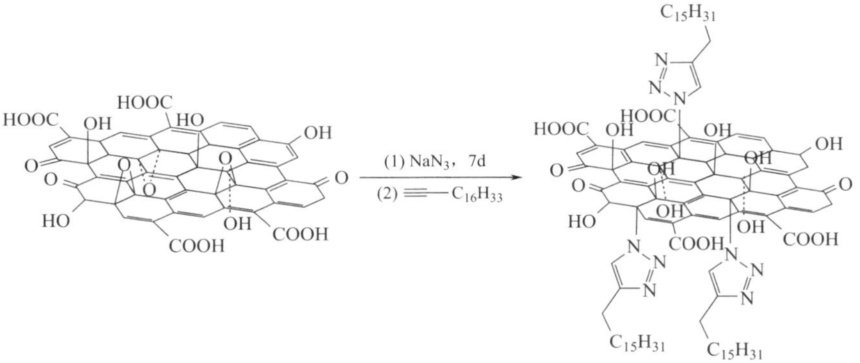

图 3.6 GO-C_{18} 的合成方法[36]

Yang 等[37] 人 2011 年报道了将 GO 表面的羟基经过酯化和取代反应，制备了表面叠氮基化 GO 的研究结果。该方法合理利用了 GO 表面丰富的羟基得到叠氮基团，增加了点击反应的活性位点，从而使得 GO 表面接枝聚合物量明显增大。研究人员将 GO 和 2-溴异丁酰溴常温下搅拌 2 天，经酯化反应后分散在 DMF 中，而后加入 NaN_3 在室温下搅拌 24 小时，得到叠氮基改性的 GO(GO-N_3)，最后与末端带炔基的 PS 反应得到 PS 修饰的 GO(GO-PS)。

3.4 非共价键修饰

非共价键主要是利用修饰成分与 GO 的氢键、π-π 作用、阳离子-π 或范德华相互作用修饰 GO。氢键、π-π 堆积、阳离子-π 或范德华这些作用主要发生在 GO 的氧化或 sp^2 碳区域，其中 π-π 相互作用的强度取决于吸附分子的 π 电子体系中芳环的数量和分子芳环与石墨材料表面之间的接触曲率。

  Liu 等[38]人利用 GO 超大的比表面积以及 GO 本身强大的 π-π 相互作用,通过吸附亚甲基蓝(Methylene Blue,简称 MB)和甲基紫(Methyl Violet,简称 MV)的实验,证明 GO 可作为环境污染物吸附剂。通过离心真空蒸发法构筑三维(Three-Dimensional,简称 3D)GO 海绵,并利用 GO 除去染料制造和纺织过程中产生的主要污染物亚甲基蓝和甲基紫。研究得到,在 3D GO 海绵上,MB 和 MV 通过 GO 强大的 π-π 作用和阴离子-阳离子相互作用完成吸附过程,二者的活化能分别为 50.3kJ/mol 和 70.9kJ/mol。石高全研究组[39]利用 GO 与小分子间的 π 共轭体系作用,合成了芘丁酸修饰的 GO。除小分子外,Qi 等[40]人将含有苯环刚性结构的三嵌断聚合段聚物(PEG-OPE)与经还原修饰过的 GO 依靠强 π-π 相互作用形成复合材料,如图 3.7 所示。

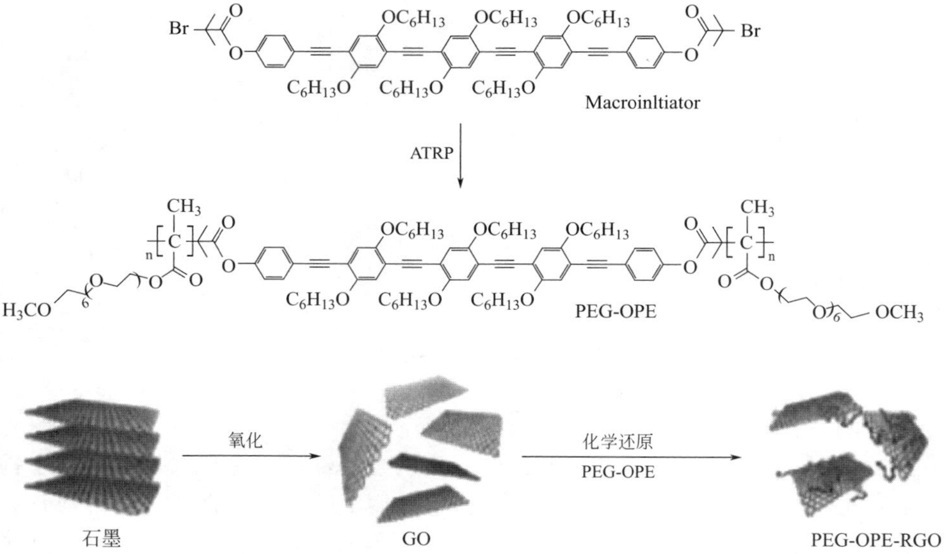

图 3.7 PEG-OPE-RGO 合成[40]

  氢键作用和范德华作用的存在是很多分子对 GO 修饰改性的基础。当功能分子中含有含氧官能团或者氨基等活性基团时,极易与 GO 上的含氧官能团形成氢键作用。Tan 等[41]人研究了如图 3.8 中链状、树枝状等大分子与 GO 氢键的作用,形成水凝胶后可以大面积构筑 GO 复合薄膜材料。磺化聚醚醚酮也用于非共价键功能化 GO。与聚偏氟乙烯相比,磺化聚醚醚酮修饰后的材料杨氏模量提高了 160%,氧气透过率降低了 91%[42]。磺化聚苯胺被用来改善 RGO 的水分散性[43],其中可能涉及 π-π 堆积和阳离子-π 相互作用,并且获得了高电导率以及良好的电催化活性和稳定性。

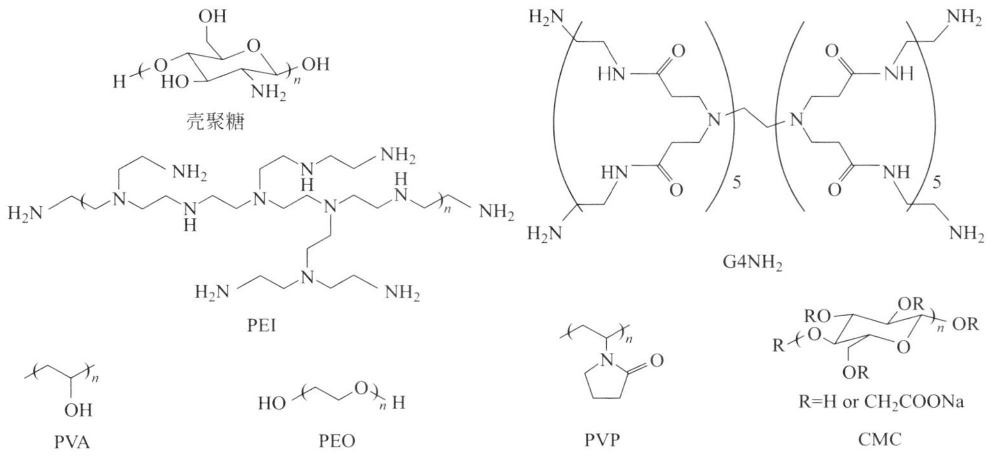

图 3.8 GO 形成氢键作用的大分子[41]

3.5 氧化石墨烯交联

GO 本身独特的二维纳米片层结构极易构筑宏观自支撑的薄膜材料。然而，GO 薄膜通常仅可保持单层 GO 片中不到 10% 的刚度和 1% 的极限拉伸强度[44]，GO 薄膜中的力学性能主要归因于堆叠的 GO 片之间较弱氢键作用和 π-π 相互作用的贡献。此外，GO 自支撑膜易受水影响，少量水都会使 GO 膜出现溶胀状态。为了充分利用 GO 膜及其复合材料的所有可能功能，经常需要 GO 片层之间形成有效交联。与前述共价与非共价键修饰作用机制类似，引入不同官能团分子与 GO 片层间发生相互作用，可实现 GO 交联并增强其机械强度等性能。

目前对于 GO 材料，由于不同作用力而交联构筑不同类型的二维、三维复合材料的研究中，经常受到生物材料结构的启发。例如，模仿类贝壳结构材料-软硬结合的特殊结构，以 GO 和官能团分子或者纳米线等进行复合可获得高强高韧等功能性材料[45]，如图 3.9 所示。

在图 3.9 中，GO 交联分为两种方式：一是以 GO 为基础的二元结合复合物组装的三维层状结构；二是包含 GO 三元结合插层结构。在二元结合中，主要是将 GO 与金属离子、有机小分子、线性聚合物、树枝状聚合物等以不同方式通过离子键、氢键、共价键等复杂的相互作用交联后得到三明治结构的层状自支撑材料。其

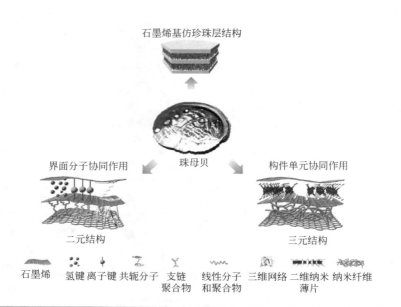

图 3.9 GO 交联的两种方式模型[46]

中，金属离子与 GO 之间交联作用得到的复合膜材料模量高但伸长率较低[46]。在二元结合的材料中，有机材料受到的关注较多，其中有机材料通常充当"软"组分，而 GO 则是该类材料中的"硬"组分。有机小分子如硼酸与 GO、10,12-二十五碳二炔-1-醇（PCDO）与 GO 交联[47,48]，都能得到机械性能优异的自支撑膜材料；有机大分子聚合物的种类繁多，主要可分为线性和树枝状聚合物两大类。线性聚合物一般是带有活性官能团的高分子，例如海藻酸钠[49]、聚乙烯醇（Polyvinyl Alcohol，简称 PVA）[50]、聚甲基丙烯酸甲酯（Polymethyl Methacrylate，简称 PMMA）[51]、聚乙烯亚胺[52]、聚丙烯胺[53] 天然生物大分子蚕丝蛋白[25] 等，其活性官能团与 GO 表面活性官能团经过复杂的交联作用，组装得到功能化的 GO 复合材料。线性聚合物的引入使得复合材料柔性增强，其伸长率明显提高。但与金属离子交联作用不一致的是，由于协同作用的影响使得材料的强度得到了提高。树枝状的聚合物相对比较少，主要是因为不同种类树枝状聚合物本身比较难以得到；Hu 等[54] 人将 GO 与超支化聚缩水甘油醚纺丝交联，得到交联程度较高的 GO 复合材料。三元插层，主要是在 GO 片层间引入三维网络、二维片层和纳米纤维等以非共价键等交联作用组装得到插层的类三明治结构材料。三元结构的目的主要是将 GO 与其他二者进行组装使得复合材料出现协同作用增强性能，如将 GO 与 A_2O_3/PVA[55]、纳米纤维素/PCDO 以及 MoS_2/TPU[56] 等三元复合均能得到性能良好的层状结构复合材料。关于 GO 组装得到的材料形貌、结构和应用等将在下一章专门介绍。

3.6 小结

本章主要介绍了 GO 的修饰与改性方法及其潜在应用。GO 易于修饰与改性的特点从很大程度上扩展了 GO 的应用领域，也是 GO 获得关注的一个重要原因。由于 GO 独特的结构，其与功能性分子复合后，复合基材很多原有性能均得以改善或者提升，使得 GO 成了非常灵活的添加剂材料或主体材料。深入了解并掌握 GO 的改性方法和技术，将有可能进一步发挥 GO 本身的特性，并使其应用范围更为广阔。在未来的研究中，更多关于 GO 修饰与改性方法与技术的探索和发掘将会持续进行。

—— 参考文献 ——

[1] Song Li，Khoerunnisa Fitri，Gao Wei，et al. Effect of high-temperature thermal treatment on the structure and adsorption properties of reduced graphene oxide [J]. Carbon，**2013**，52（2）：608-612.

[2] Longo Angela，Ambrosone G，Coscia U，et al. Fluorimetric Study of Graphene Oxide Reduction by Microwave Heating [C]. Photonic Technologies. IET，2017.

[3] Graeme Williams，Brian Seger，Kamat Prashant V. TiO_2-graphene nanocomposites. UV-assisted photocatalytic reduction of graphene oxide [J]. Acs Nano，**2008**，2（7）：1487-1491.

[4] Vasilios Georgakilas，Michal Otyepka，Bourlinos Athanasios B，et al. Functionalization of graphene：covalent and non-covalent approaches，derivatives and applications [J]. Chemical Reviews，**2012**，112（11）：6156-6214.

[5] Dreyer Daniel R.，Todd Alexander D.，Bielawski Christopher W.. Harnessing the chemistry of graphene oxide [J]. Chemical Society Reviews，**2014**，43（15）：5288-5301.

[6] Liu Zhuang，T Robinson Joshua，Sun Xiaoming，et al. PEGylated nanographene oxide for delivery of water-insoluble cancer drugs [J]. Journal of the American Chemical Society，**2008**，130（33）：10876-10877.

[7] Monica Veca L，Lu Fushen，J Meziani Mohammed，et al. Polymer functionalization and solubilization of carbon nanosheets [J]. Chemical Communications，**2009**，45（18）：2565-2567.

[8] Stankovich Sasha，Piner Richard D.，Nguyen Son Binh T.，et al. Synthesis and exfoliation of isocyanate-treated graphene oxide nanoplatelets [J]. Carbon，**2006**，44（15）：3342-3347.

[9] Xu Yanfei，Liu Zhibo，Zhang Xiaoliang，et al. A Graphene Hybrid Material Covalently Functionalized with Porphyrin：Synthesis and Optical Limiting Property [J]. Advanced Materials，**2010**，21（12）：1275-1279.

[10] Hu Xiangang，Mu Li，Wen Jianping，et al. Covalently synthesized graphene oxide-aptamer nanosheets for efficient visible-light photocatalysis of nucleic acids and proteins of viruses [J]. Carbon，**2012**，50（8）：2772-2781.

[11] Mejias Carpio Isis，Mangadlao Joey，Nguyen Hang，et al. Graphene oxide functionalized with ethylenediamine triacetic acid for heavy metal adsorption and anti-microbial applications [J]. Carbon，**2014**，77

(10): 289-301.

[12] Wu Huixia, Shi Haili, Wang Yapei, et al. Hyaluronic acid conjugated graphene oxide for targeted drug delivery [J]. Carbon, **2014**, 69 (69): 379-389.

[13] Li Zhe Fei, Zhang Hangyu, Liu Qi, et al. Covalently-grafted polyaniline on graphene oxide sheets for high performance electrochemical supercapacitors [J]. Carbon, **2014**, 71 (5): 257-267.

[14] Yang Yongfang, Wang Jie, Zhang Jian, et al. Exfoliated graphite oxide decorated by PDMAEMA chains and polymer particles [J]. Langmuir the Acs Journal of Surfaces Colloids, **2009**, 25 (19): 11808-11814.

[15] Wan Yan Jun, Tang Long Cheng, Gong Li Xiu, et al. Grafting of epoxy chains onto graphene oxide for epoxy composites with improved mechanical and thermal properties [J]. Carbon, **2014**, 69 (2): 467-480.

[16] Dreyer Daniel R., Jarvis Karalee A., Ferreira Paulo J., et al. Graphite Oxide as a Dehydrative Polymerization Catalyst: A One-Step Synthesis of Carbon-Reinforced Poly (phenylene methylene) Composites [J]. Macromolecules, **2015**, 44 (19): 7659-7667.

[17] Jahandideh Sara, Javad Sarraf Shirazi Mohammad, Tavakoli Mitra. Mechanical and thermal properties of octadecylamine-functionalized graphene oxide reinforced epoxy nanocomposites [J]. Fibers and Polymers, **2017**, 18 (10): 1995-2004.

[18] Chen Jin Long, Yan Xiu Ping. A dehydration and stabilizer-free approach to production of stable water dispersions of graphene nanosheets [J]. Journal of Materials Chemistry, **2010**, 20 (21): 4328-4332.

[19] Liu Jianghua, Chen Guosong, Jiang Ming. Supramolecular Hybrid Hydrogels from Noncovalently Functionalized Graphene with Block Copolymers [J]. Macromolecules, **2017**, 44 (19): 7682-7691.

[20] Yang Huafeng, Shan Changsheng, Li Fenghua, et al. Covalent functionalization of polydisperse chemically-converted graphene sheets with amine-terminated ionic liquid [J]. Chemical Communications, **2009**, 45 (26): 3880-3882.

[21] Liu Chengbao, Qiu Shihui, Du Peng, et al. An ionic liquid-graphene oxide hybrid nanomaterial: synthesis and anticorrosive applications [J]. Nanoscale, **2018**, 10 (17): 8115-8124.

[22] Zhu Shengming, Dong Xufeng, Gao Song, et al. Uniformly Grafting SnO_2 Nanoparticles on Ionic Liquid Reduced Graphene Oxide Sheets for High Lithium Storage [J]. Advanced Materials Interfaces, **2018**, 5 (9): 1701685.

[23] Wang Yanying, Li Chunya, Wu Tsunghsueh, et al. Polymerized ionic liquid functionalized graphene oxide nanosheets as a sensitive platform for bisphenol A sensing [J]. Carbon, **2018**, 129: 21-28.

[24] An Zhi, Compton Owen, W Putz Karl, et al. Bio-inspired borate cross-linking in ultra-stiff graphene oxide thin films [J]. Advanced Materials, **2011**, 23 (33): 3842-3846.

[25] Park Sungjin, Dikin Dmitriy A., Nguyen Son Binh T., et al. Graphene Oxide Sheets Chemically Cross-Linked by Polyallylamine [J]. Journal of Physical Chemistry C, **2009**, 113 (36): 15801-15804.

[26] Xu Zhiwei, Zhang Jiguo, Shan Mingjing, et al. Organosilane-functionalized graphene oxide for enhanced antifouling and mechanical properties of polyvinylidene fluoride ultrafiltration membranes [J]. Journal of Membrane Science, **2014**, 458 (10): 1-13.

[27] Gonçalves Gil, Marques Paula A. A. P, Barros-Timmons Ana, et al. Graphene oxide modified with PMMA via ATRP as a reinforcement filler [J]. Journal of Materials Chemistry, **2010**, 20 (44): 9927-9934.

[28] Mcgrail Brendan T., Rodier Bradley J., Pentzer Emily. Rapid Functionalization of Graphene Oxide in Water [J]. Chemistry of Materials, **2014**, 26 (19): 5806-5811.

[29] Fang Ming, Wang Kaigang, Lu Hongbin, et al. Covalent polymer functionalization of graphene nanosheets and mechanical properties of composites [J]. Journal of Materials Chemistry, **2009**, 19 (38): 7098-7105.

[30] Huang Yingjuan, Qin Yawei, Zhou Yong, et al. Polypropylene/Graphene Oxide Nanocomposites Prepared by In Situ Ziegler-Natta Polymerization [J]. Chemistry of Materials, **2010**, 22 (13): 4096-4102.

[31] Hwa Lee Sun, R Dreyer Daniel, An Jinho, et al. Polymer Brushes via Controlled, Surface-Initiated Atom Transfer Radical Polymerization (ATRP) from Graphene Oxide [J]. Macromolecular Rapid Communications, **2010**, 31 (3): 281-288.

[32] Ma Lijun, Yang Xiaoming, Gao Lingfeng, et al. Synthesis and characterization of polymer grafted graphene oxide sheets using a Ce (IV) /HNO$_3$ redox system in an aqueous solution [J]. Carbon, **2013**, 53 (Complete): 269-276.

[33] Sun Shengtong, Cao Yewen, Feng Jiachun, et al. Click chemistry as a route for the immobilization of well-defined polystyrene onto graphene sheets. J Mater Chem [J]. Journal of Materials Chemistry, **2010**, 20 (27): 5605-5607.

[34] Kou Liang, He Hongkun, Gao Chao. Click chemistry approach to functionalize two-dimensional macromolecules of graphene oxide nanosheets [J]. Nano-Micro Letters, **2010**, 2 (3): 177-183.

[35] Pan Yongzheng, Bao Hongqian, Sahoo Nanda Gopal, et al. Water-Soluble Poly (N-isopropylacrylamide)-Graphene Sheets Synthesized via Click Chemistry for Drug Delivery [J]. Advanced Functional Materials, **2011**, 21 (14): 2754-2763.

[36] Riccardo Salvio, Sven Krabbenborg, Naber Wouter J M, et al. The formation of large-area conducting graphene-like platelets [J]. Chemistry **2010**, 15 (33): 8235-8240.

[37] Yang Xiaoming, Ma Lijun, Wang Sheng, et al. "Clicking" graphite oxide sheets with well-defined polystyrenes: A new Strategy to control the layer thickness [J]. Polymer, **2011**, 52 (14): 3046-3052.

[38] Liu Fei, Chung Soyi, Oh Gahee, et al. Three-dimensional graphene oxide nanostructure for fast and efficient water-soluble dye removal [J]. Acs Applied Materials Interfaces, **2012**, 4 (2): 922-927.

[39] Xu Yuxi, Hua Bai, Lu Gewu, et al. Flexible graphene films via the filtration of water-soluble noncovalent functionalized graphene sheets [J]. Journal of the American Chemical Society, **2008**, 130 (18): 5856-5857.

[40] Qi Xiaoying, Pu Kan Yi, Li Hai, et al. Amphiphilic Graphene Composites &-dagger [J]. Angewandte Chemie International Edition, **2010**, 122 (49): 9616-9619.

[41] Tan Zhibing, Zhang Miao, Li Chun, et al. A General Route to Robust Nacre-Like Graphene Oxide Films [J]. Acs Applied Materials Interfaces, **2015**, 7 (27): 15010-15016.

[42] Layek Rama K., Das Ashok Kumar, Min Jun Park, et al. Enhancement of physical, mechanical, and gas barrier properties in noncovalently functionalized graphene oxide/poly (vinylidene fluoride) composites [J]. Carbon, **2015**, 81 (1): 329-338.

[43] Hua Bai, Xu Yuxi, Zhao Lu, et al. Non-covalent functionalization of graphene sheets by sulfonated polyaniline [J]. Chemical Communications, **2009**, 13 (13): 1667-1669.

[44] Paci Jeffrey T., Belytschko Ted, Schatz George C.. Computational Studies of the Structure, Behavior upon Heating, and Mechanical Properties of Graphite Oxide [J]. J. phys. chem. c, **2007**, 111 (49): 18099-18111.

[45] Wan Sijie, Peng Jingsong, Li Yuchen, et al. Use of Synergistic Interactions to Fabricate Strong, Tough, and Conductive Artificial Nacre Based on Graphene Oxide and Chitosan [J]. Acs Nano, **2015**, 9 (10): 9830-9836.

[46] Cheng Qunfeng, Duan Jianli, Zhang Qi, et al. Learning from nature: constructing integrated graphene-based artificial nacre [J]. Acs Nano, **2015**, 9 (3): 2231-2234.

[47] Sungjin Park, Kyoung-Seok Lee, Gulay Bozoklu, et al. Graphene oxide papers modified by divalent ions-enhancing mechanical properties via chemical cross-linking [J]. Acs Nano, **2008**, 2 (3): 572-578.

[48] Cheng Qunfeng, Wu Mengxi, Li Mingzhu, et al. Ultratough artificial nacre based on conjugated cross-linked graphene oxide [J]. Angew Chem Int Ed Engl, **2013**, 125 (13): 3750-3755.

[49] Lee Hanleem, Han Guebum, Kim Meeree, et al. High mechanical and tribological stability of an elastic ultrathin overcoating layer for flexible silver nanowire films [J]. Advanced Materials, **2015**, 27 (13): 2252-2259.

[50] Li Yuan-Qing, Yu Ting, Yang Tian-Yi, et al. Bio-inspired nacre-like composite films based on graphene with superior mechanical, electrical, and biocompatible properties [J]. Advanced Materials, **2012**, 24 (25): 3426-3431.

[51] Xu Yuxi, Hong Wenjing, Bai Hua, et al. Strong and ductile poly (vinyl alcohol) /graphene oxide composite films with a layered structure [J]. Carbon, **2009**, 47 (15): 3538-3543.

[52] Tian Ye, Cao Yewen, Wang Yu, et al. Realizing ultrahigh modulus and high strength of macroscopic graphene oxide papers through crosslinking of mussel-inspired polymers [J]. Advanced Materials, **2013**, 25 (21): 2980-2983.

[53] You-Hao Yang, Laura Bolling, Priolo Morgan A, et al. Super gas barrier and selectivity of graphene oxide-polymer multilayer thin films [J]. Advanced Materials, **2013**, 25 (4): 503-508.

[54] Hu Xiaozhen, xu Zhen, Gao Chao. Multifunctional, supramolecular, continuous artificial nacre fibres [J]. Scientific Reports, **2012**, 2 (10): 767.

[55] Wang Jinrong, Qiao Jinliang, Jianfeng Wang, et al. Bioinspired Hierarchical Alumina-Graphene Oxide-Poly (vinyl alcohol) Artificial Nacre with Optimized Strength and Toughness [J]. Acs Appl Mater Interfaces, **1944**, 7 (17): 9281-9286.

[56] Gong Shanshan, Cui Wei, Zhang Qi, et al. Integrated Ternary Bioinspired Nanocomposites via Synergistic Toughening of Reduced Graphene Oxide and Double-Walled Carbon Nanotubes [J]. Acs Nano, **2015**, 9 (12): 11568-11573.

第4章
CHAPTER 4

氧化石墨烯组装

石墨烯或 GO 很多独特性质和与纳米尺寸紧密相关，将这些纳米尺度上的性质无损或者以较高效率转移到宏观尺寸是石墨烯应用中的重要研究课题。因此，将石墨烯纳米片层结构组装成易于应用并具有优异性能的宏观结构至关重要。目前，对 GO 已经实现了不同维度的组装，构建了包括纤维（一维）、薄膜（二维）、泡沫或块体材料（三维）等众多形态各异的碳基功能材料，而构建这些不同维度 GO 基宏观材料的方法也非常丰富。在组装中主要利用的相互作用是分子间作用力、官能团辅助成键和空间互补效应（如静电力、氢键、范德华力、官能团的电子效应等），在溶液中或两相界面处缔结成热力学稳定、结构确定、性能特殊的聚集体[1,2]。

本章将主要介绍以 GO 为主体材料进行的不同维度组装的制备方法、性能及应用。

4.1 氧化石墨烯纤维（一维）组装及应用

常规合成纤维方法包括熔体纺丝和溶液纺丝。熔体纺丝适用于熔融过程不会明显分解的聚合物纺丝。对于在熔点附近温度会严重分解的聚合物，优选将聚合物溶解在溶剂中进行溶液纺丝。溶液纺丝法包括湿法纺丝、干喷-湿法纺丝和干法纺丝。最近利用这几种纺丝技术制造 GO 纤维的研究均有报道[3~5]。由于 GO 具有两亲性，易于分散到水中，所以溶液纺丝法成为制备 GO 纤维的主要方法。此外，除了这些传统的溶液纺丝方法之外，最新开发的电泳法和薄膜转化法也可以用于制造

GO 纤维[6,7]。

4.1.1 氧化石墨烯纤维的制备及性能研究

（1）湿法纺丝

湿法纺丝的典型过程为：首先将 GO 片分散到水溶液中得到稳定的 GO 悬浮液，然后将 GO 悬浮液注入凝固浴中以形成凝胶状 GO 纤维，凝固一段时间后，再通过提取凝胶状纤维然后干燥来获得 GO 纤维。根据需要，GO 纤维还可以进行还原，制备石墨烯纤维。在此过程中，凝胶状 GO 纤维的纺丝原液、凝固浴和移动速度都会影响 GO 纤维的均匀性和连续性[5]。纤维的制造流程如图 4.1(a) 和（b）所示：

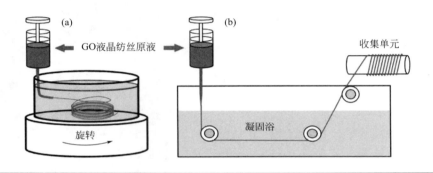

图 4.1 通过旋转凝固浴（a）和收集单元（b）进行湿法纺丝合成纤维示意[5]

Xu 等[8]人报道了通过湿法纺丝工艺制备不添加聚合物等助剂的石墨烯纤维。该研究通过湿法拉伸和离子交联实现 GO 液晶纺制，再经过化学还原后获得高强度、高导电性的石墨烯纤维，这种石墨烯纤维可以应用于功能性纺织品、柔性和可穿戴传感器以及超级电容器设备领域。

此外，许多工作还研究了 GO 纤维湿法纺丝过程中的影响因素。

Jalili 等[9]人研究解决了 GO 液晶分散体形成中的关键问题，验证了湿法纺丝制备长尺寸微米级纤维和纱线的可加工性。作者通过流变学研究了 GO 片材的尺寸、分散体的浓度和液晶性之间的关系。研究表明，液晶态 GO 分散体表现出非牛顿液体的流变行为，流动曲线显示屈服值（产生黏性流动所需的最低剪切应力）类似于塑料流体的流变行为。当 GO 浓度高于 0.75mg/mL 时，纺丝溶液的内聚力显著提高，能够产生很长的坚固凝胶纤维，纤维在随后干燥的过程中也没有破裂。值得注意的是，可纺性液晶态 GO 分散体的浓度和黏度可分别低至约 0.75mg/mL 和 30Pa·s，该浓度比先前报道的液晶态 GO 分散体湿纺丝的值低约两个数量级。研究认为，较大的 GO 片尺寸和长径比（平均横向尺寸约 37μm，厚度 0.8nm，平均长径比约为

45000）有助于 GO 片在纤维内定向排列。此外，该研究在碱性凝固浴（即 NaOH 或 KOH）中通过一步连续纺丝法原位纺织液晶态 GO 来制备石墨烯纤维和纱线，不需要后处理工艺。研究发现，这种方法制备的石墨烯纤维的导热系数为 1435W/m·K，远高于碳纳米管纤维（380W/m·K）[10]、石墨烯纸（112W/m·K）[11] 以及多晶石墨材料（200W/m·K）[12]。

Nuray 等[3] 人研究了时间、剥离温度和 GO 的分散方法等因素对湿法纺丝制备的连续 GO 纤维性能的影响。研究表明，在剥离状态、分散状态和纤维状态时，晶体结构对参数变化的响应是不同的。例如，GO 纤维的结晶尺寸和层数是大于 GO 分散液的，而 GO 纤维的结晶度则小于 GO 分散液。剥离时间延长会导致纤维的特克斯数增大，然而，相比于短时超声和机械均化，较长的超声时间导致特克斯数较低。与机械均化相比，更短的超声处理和更短的剥离时间导致更高的电导率。然而，研究还表明，较短的超声处理时间可能会导致纤维断裂强度增加。

Eom 等[13] 人研究了在 GO 纤维凝结过程中，多价金属阳离子对 GO 纤维结构的影响。该研究中纤维的具体制备方法为：将 GO 分散液通过喷丝头［直径 400μm］以固定速度（10mL/h）纺入凝固浴［氯化钴（Ⅱ），氯化铝（Ⅲ）或三价铁（Ⅲ）氯化物水溶液］中。将在凝固浴中纺制的 GO 凝胶纤维连续地收集在卷轴上，无牵伸。GO 纤维在凝固浴中的停留时间约为 6s。然后，将 GO 纤维用去离子水洗涤并在室温下恒湿干燥。研究发现，通过三价盐凝固浴纺制的纤维结构很紧密，而通过二价盐凝固浴纺制的纤维外观出现了扭曲变形，如图 4.2 所示。这种结构差异表明，GO 凝聚过程中阳离子的类型是影响 GO 纤维形态和性质的主要因素。与二价阳离子相比，三价阳离子的快速扩散可以诱导 GO 纤维形成高度致密的结构。此外，三价金属阳离子（Fe^{3+}）凝固浴制备的 GO 纤维机械性能更优异，测得该凝固浴制备的纤维拉伸强度为 486MPa，杨氏模量为 81GPa，断裂伸长率为 0.62%，韧度为 1.47MJ/m^3。扩散到 GO 凝胶内部结构中的阳离子与 GO 中含氧官能团所带的负电荷发生静电吸引，导致基面黏结，而三价阳离子与纤维的界面结合作用更强，形成的纤维结构更致密，所以能够提高 GO 纤维的强度。

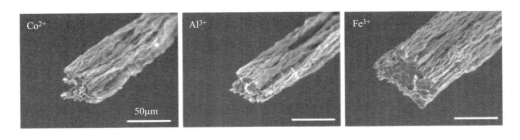

图 4.2　不同凝固浴中纺织的 GO 纤维的微观形态[13]

Zhang 等[14] 人报道了使用离子液体作为凝结剂通过湿纺技术制造高性能 GO 纤维的方法。结果显示，通过选择具有设计结构的阳离子的离子液体来调节 GO 片层的官能团和离子液体阳离子之间的相互作用，能够产生具有不同机械性能的 GO 纤维。由于不涉及有机溶剂或后拉伸工艺，所以该工艺更加环保和简便。具体制备方法为：用旋转针（内径 0.5mm）将 GO 水性分散液（13mg/mL）装入塑料注射器中，并用注射泵注入离子液体凝固浴（0.005mol/L，离子液体/水）中，速率为 0.4mL/min。GO 溶液立即凝结形成 GO 凝胶纤维。为了进行拉伸测试，将 GO 纤维悬浮在两根平行棒上并在室温下干燥，使用氢碘酸（40%）在 80℃下过夜还原 GO 纤维，选择的凝结剂分别为 1-氨丙基-3-甲基咪唑鎓溴化物（[APMIm]$^+$[Br]$^-$）、1-丁基-3-甲基咪唑鎓乳酸盐（[BMIm]$^+$[LA]$^-$）和 1-丁基-1-甲基吡咯烷鎓溴化物（[BMPr]$^+$[Br]$^-$），它们的阳离子具有相似的分子大小但结构不同。通过 [BMIm]$^+$[LA]$^-$凝固浴制备的 GOF-BMIm 纤维的平均杨氏模量为 32.8GPa，平均拉伸强度为 387.3MPa。对于 GOF-BMPr 纤维，平均杨氏模量和平均拉伸强度分别为 16.9GPa 和 304.1MPa。相比于以前报道的使用其他种类凝结剂制备的 GO 纤维，这两种 GO 纤维显示出很好的机械性能。类似于铵基团（-NH$_3^+$），[BMIm]$^+$ 和 [BMPr]$^+$ 阳离子可以与 GO 层上的酸性基团（面内羟基和边缘连接的羧基）形成离子键。GO 层之间的弱静电排斥有利于 GO 层组装成纤维形状。由于离子液体阳离子和 GO 层之间的强电吸引力，与先前报道的 GO 纤维相比，制备的 GO 纤维中的 GO 片显示出更高的取向度和填充度，而不需要添加有机溶剂作为共凝结剂和经过拉伸过程。

Cao 等[15] 人利用 3D 打印技术制造了具有优异机械性能的石墨烯纤维基网络结构。该技术是沿可编程轨道从可移动喷丝头直接纺制 GO 分散体，由此产生的 GO 网络结构可以成功转化为具有更好机械性能的 RGO 网络结构。纤维纺丝加工过程中形成的原位焊接接头使得所得到的网络具有整体结构，并且纺出的纤维及纤维网的机械性能高、导电性好。该研究的制备工艺如图 4.3 所示。

(2) 干法纺丝

干法纺丝也使用 GO 分散液（主要分散在水中）作为纺丝原液，但是不再使用凝固浴[16,17]。主要方法是：将 GO 分散液注入并密封在管道中，通过在高温下加热或化学还原后，管道中沉淀出凝胶状的纤维，然后除去溶剂得到干燥的 RGO 纤维。这里的 GO 分散液是指具有大尺寸 GO 的胶体（液溶胶）（平均 GO 片径远大于 100nm）。一方面，高温下 GO 分散液快速移动，破坏由 Zeta（ζ）电位维持的平衡，从而增加 GO 片材碰撞和沉淀的可能性。另一方面，高温或化学还原会去除 GO 中的含氧基团，从而降低 GO 分散体的绝对 ζ 电位值。由于缺乏足够的静电排斥，GO 片沉淀并组装成纤维。

Dong[18]、Yu[19] 和 Li 等[20] 人的研究表明，在凝固浴中沉淀的凝胶状纤维被溶剂溶胀，经过干燥后，纤维直径变小（约下降 80%）。GO 片在 220~230℃加

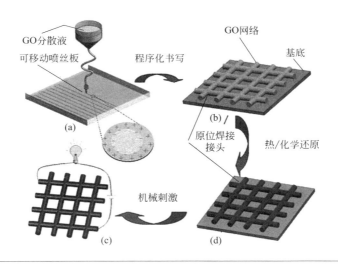

图 4.3　纤维及纤维网络的制备工艺[15]

热过程中被沉淀出来,这实际上是一种溶剂热过程(当水用作 GO 分散溶剂时为水热过程),并且通过管状的成型模具将 GO 片材限制在一维空间,所以得到的 GO 宏观体材料形态为纤维。实际上这种方法制得的是 RGO 纤维,是因为在沉淀的过程中,高温会将 GO 热还原为 RGO。据报道,加热温度为 180℃时可以除去 GO 中 27%的氧[21],并且在 200℃下部分羟基、环氧基和羧基开始被还原。因此,无须通过后处理,通过水热过程合成的 GO 纤维的电导率(约 10S/cm)也较高[18,22,23]。Yu 等[19] 人的研究还表明,制备得到的 RGO 纤维中有一定程度的 RGO 片层取向,这归因于毛细管引起的剪切力和表面张力引起的片层的相互作用。

Li 等[20] 人报道了在管式模具中通过化学还原 GO 制备 RGO 纤维的研究。研究人员将 GO 和维生素 C 溶液注入 PP 管中,然后 80℃下加热 1 小时,GO 被还原的同时也被组装成凝胶状态 RGO 纤维,干燥后纤维直径减少 95%~97%,测得 RGO 纤维的极限断裂伸长率为 20%,拉伸强度为 150MPa,电导率约为 8S/cm。此外,Fan 等[24] 人研究了改变模具结构和纺丝原液的组成,发现通过干法纺丝也可以制造中空和复合 GO 纤维。

(3) 干喷-湿法纺丝

干喷-湿法纺丝是合成纤维的一种重要的传统纺丝方法,这种方法可以用来纺制具有均匀结构和圆形横截面的纤维。如图 4.4 所示,Xiang 等[17] 人将 GO 纳米片(GONRs)分散在氯磺酸中形成高浓度的各向异性液晶相,然后将液晶溶液注入乙醚中制备 GO 纤维。在 1500℃下对初纺的 GONR 纤维进行热处理,产生热还原的石墨烯纳米带(Grapheme Nanoribbons,简称 trGNR)纤维,其拉伸强度为

378MPa，杨氏模量为 36.2GPa，电导率为 285S/cm。由于通过气隙纺丝和预拉伸退火，纤维内部可产生更有序的分子排列，所以 trGNR 纤维的性能更好。图 4.5 为干喷-湿法纺丝 GO 纤维的表面 (a)、(c) 和横截面 (b) 图像，表明选择适当的溶剂和凝固剂（氯磺酸和乙醚），干喷-湿法纺丝可以制备具有光滑表面和圆形横截面形状的石墨烯纤维。

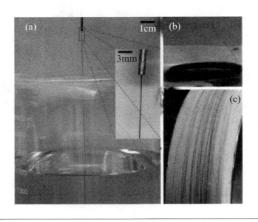

图 4.4 RGO 纤维的纺织过程[17]

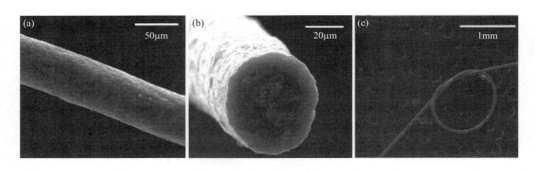

图 4.5 RGO 纤维的表面和横截面图像[17]

Seyedin 等[25]人报道了使用纺织机械编织高柔韧性的 GO 纤维。这项工作中 GO 纤维通过干喷湿法纺丝方法制造，得到的 GO 纤维的杨氏模量约为 7.9GPa，拉伸强度约为 135.8MPa，断裂应变约为 5.9%，韧性约为 5.7MJ/m^3。结合合适的纺丝方法、凝固浴和纺丝条件，可以制备具有较高韧性的 GO 纤维，这是他们成功编织的关键因素。GO 分散液为全向列相液晶材料，这一典型特征赋予了 GO 分散液可纺性。凝固浴的溶剂组成和纤维纺丝结构在实现可纺性方面也起到重要作用。例如，当在湿纺过程中使用 $CaCl_2$ 水溶液作为凝固浴时，GO 纤维漂浮在凝固浴中，这使得收集纤维相当困难，并且阻碍了纤维的进一步加工。在干喷湿纺过程中，GO 喷射（即 GO 在空气中的分散流体）不能穿透凝固浴，这不利于形成初始纤维，而使用乙醇和异丙醇作为 $CaCl_2$ 的溶剂效率也较低，使用有机溶剂还会影响纤维的形貌和力学性能。例如，使用10%（质量）的氯化钙乙醇溶液作为凝固浴，得到的 GO 纤维的横截面不规则，并且机械性能较差。当使用的溶剂为乙醇/水混合物（50/50 体积比）时，GO 纤维的横截面变得规则（即接近圆形）并且机械性能得到改善。此外，纺制过程中的凝固速率也是非常重要的，凝固速率快会导致纤维横截面不规则和形成多孔形态，这两者都将导致机械性能变差。另一方面，乙醇/水（50/50 体积比）的混合物具有低的混凝率，由此可以获得良好的可加工性，使纤维圆度提高并形成致密的结构。所以，通过控制凝固速率，可以制备具有接近圆形横截面的 GO 纤维。研究发现，通过湿法纺丝制备的 GO 纤维的横截面不规则，并且是高度多孔的，通过干喷湿纺生产的纤维具有更好的圆形度和堆积结构（即具有更少的空隙）。使用低浓度 GO 或小尺寸喷丝头生产的纤维具有较高的杨氏模量和拉伸强度，这是因为沿着纤维长度的结构缺陷减少。随着 GO 纤维直径减小，GO 片材和有序的液晶态 GO 区域的数量也减少，因此，每根纤维长度的缺陷变小。于是，GO 片材和有序的液晶态 GO 区域之间的交联和范德华相互作用也变得更低，导致较低的滑移强度并因此降低了断裂应变和韧性。这项工作突出了 GO 纤维作为一种新型纺织品的应用潜力。

（4）电泳组装法

氧化石墨片层表面或边缘含有的羧基等含氧官能团，可在水溶液中发生电离而带负电荷，因此在电场的作用下移动，同时 GO 分散体被认为是胶体，所以电泳自组装法可以用于制备 GO 纤维。Yun 等[26]人在不使用任何聚合物或表面活性剂的情况下，通过电泳自组装方法制得了 RGO 纤维。具体的制备方法为：将石墨尖端作为正极浸入 GO 分散液中，在恒定的电位下，通过石墨尖端的退出，可使电极末端形成 GO 凝胶状纤维，再经过干燥和热还原，得到具有光滑表面和圆形横截面的 RGO 纤维。由于石墨电极的移动速度较慢（仅为 0.1mm/min），所以导致纤维产率很低，生产 1m 长的纤维大约需要一周的时间。

（5）薄膜转化法

上述四种方法的制备过程均是由 GO 溶液转化为 GO 纤维或 RGO 纤维，而薄膜转化法是利用固态 GO 或 GO 薄膜来制备 GO 纤维，其中 GO 薄膜可以直接缠绕或扭曲成 GO 纤维，所以这种方法中 GO 薄膜的制备是关键。目前，GO 溶液蒸发法和解链 CNTs 这两种方法可以用于制备 GO 薄膜。Cruz-Silva 等[27]人通过在聚四氟乙烯基板上蒸发 GO 溶液的溶剂制备了连续的 GO 膜（例如 800~1200cm^2 或更大）。GO 膜被进一步切割成长条并捻成纤维，GO 纤维的断裂伸长率高达76%，韧性高达 17J/m³。GO 纤维的横截面光滑、均匀，如图 4.6(a) 所示。将 GO 纤维在 2000℃ 热还原后，可以制备得到在室温下电导率为 416S/cm 的高导电性石墨烯基纤维。第二种方法是解链 CNTs。如图 4.6(b) 所示，Carretero-González 等[28]人将 CNTs 在 $KMnO_4/H_2SO_4$ 溶液中化学解链后转化成 GONRs，并且 GONRs 保持相互连接，在溶液中形成连续的 GONR 膜。当提取到空气中时，GONR 薄膜收缩成凝胶状纤维，最后在溶剂蒸发后形成干燥纤维，纤维经过化学和热处理后，室温导电率达到 100S·cm²/g（900℃ 热处理后）。

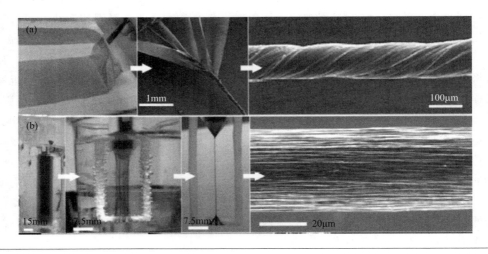

图 4.6 RGO 纤维的制备及 SEM 表征[27,28]

4.1.2 氧化石墨烯纤维的主要应用

GO 纤维经还原后具有优异的导电、导热以及机械性能。目前，已经开发出 RGO 在导线、纺织品、储能装置、智能执行器、场发射器、固相微萃取、催化等领域的应用[29]。

GO 纤维经还原处理后具有高电导率，可以用作电路中的导电线。例如，Kou 等[30]人通过连续湿纺组装技术制备了 PVA 涂覆的石墨烯纤维。研究表明，该纤维的最大拉伸应力可以达到约 200MPa，并且纤维在浸入氢碘酸后具有高导电性（约 350S/m），能够连接电路照亮发光二极管芯片。Dong[18] 将 RGO 纤维织入导电纺织品中制成智能纺织品，并用聚二甲基硅氧烷基体嵌入纺织品，该基体表现出优异的变形耐受性。

GO 纤维由于具有优异的导电性和高表面积也被用于能源领域。Aboutalebi 等[31] 人利用 GO 液晶分散体的自组装行为，制造了具有较高力学性能（杨氏模量超过29GPa)、高电导率［(2508±632)S/m］和高比表面积（2210m^2/g）的GO纤维和RGO纤维，使用RGO纤维制备的电极电容为409F/g。

Fan等[32]人通过使用离子液体（1,2-乙二胺三氟甲磺酸盐）作为凝聚浴、功能性二胺作为交联剂来连接GO片层制备了一维石墨烯纤维。经过2800℃的超高温热处理后的纤维具有优异的机械性能和电学性能，纤维的拉伸强度达到约729MPa。将石墨烯纤维浸入PVA/H$_2$SO$_4$电解质溶液中制备了准固态微型超级电容器，该电容器的体积容量高达225F/cm（在103.5mA/cm下，三电极电池中测试），循环寿命长达2000次。这些研究结果表明，石墨烯纤维未来可以替代用于小型化便携式电子应用的能量存储装置。

为解决石墨烯纤维超级电容器（Graphene Fiber Supercapacitor，简称GFSC）中面临的石墨烯片严重堆积、石墨烯纤维的疏水性和复杂制备过程等问题，Meng等[33]人开发了一种通过简单的等离子体处理提高全固态GFSC的电化学性能的方法。具体的制备方法为：首先通过湿法纺丝制备GO纤维，再将GO纤维高温（800℃）还原为石墨烯纤维（Graphene Fibers，简称GFs），将GFs在自制玻璃支撑物上拉直，该支撑物放置在功率为20W的等离子体室中间，使得等离子体可以全面处理纤维。在正常环境条件下，用空气等离子体处理GF（持续时间分别为10秒、20秒和40秒以及1分钟、2分钟、3分钟和5分钟）。再将以上过程制备的GF电极分别与PVA/H$_2$SO$_4$凝胶电解质和PVDF-共-六氟丙烯/1-乙基-3-甲基咪唑四氟硼酸盐（EMIMBF$_4$）凝胶电解质复合来制备GFSC。研究发现，超快速等离子体处理的作用是通过同时调节高微孔率的孔径分布和增强GF电极的亲水性来提高GFSC的电化学性能。在环境条件下经等离子体处理1分钟后的GFSC与未处理制备的GFSC相比，比电容提高33.1%（为36.25mF/cm^2）。在PVDF共-六氟丙烯/EMIMBF$_4$电解质中，能量密度达到18.12μW·h/cm^2，是所制备的PVA/H$_2$SO$_4$凝胶电解质中的能量密度（0.80μW·h/cm^2）的22倍。此外，经过等离子体处理的GFSC还具有优异的循环稳定性（20000次循环后，经1分钟等离子体处理后GFSC的电容保留率为96.14%）。这种等离子体处理方法未来还可以扩展到大规模制造高性能石墨烯和碳纳米管基超级电容器中。

GO纤维长径比较大、载流子密度高和导电率高，可以应用于电子发射指示器。Cruz-Silva等[27]人将GO纤维热还原后制备的石墨烯纤维用于在低导通电压下工作的电子场发射器。该石墨烯纤维在0.48V/μm时具有低导通电压，RGO片材沿着纤维轴线的取向并且这些片材边缘的拓扑缺陷可以作为有效的发射位点，使其可以产生优异的场发射性能。

Cheng等[34]人利用石墨烯片沿石墨烯氧化物纤维的螺旋排列开发了一种水分驱动旋转电机，在湿度交替状态下，其旋转速度高达5200r/min，可以被用于制造湿度开关和湿气触发式发电机。

GO纤维的大表面积和强π-π静电叠加性能使其可作为良好的固相微萃取吸附

剂。Fan 等[24]人使用 RGO 纤维分离多环芳烃（PAHs），结果表明，在 40℃ 和 20% 的盐浓度下，平衡吸收时间约为 50 分钟，在 160 次重复萃取循环之后也没有观察到提取效率的劣化。

Javier 等[28]人的研究表明，RGO 纤维还可以用于催化领域，相较于未掺杂的 RGO 纤维和氮掺杂的多壁碳纳米管，氮掺杂的 RGO 纤维对氧的电催化活性更高。

4.2 氧化石墨烯薄膜结构组装及应用

薄膜是材料的一种重要利用形式，将 GO 组装成二维薄膜是该领域中研究最早、报道最多的一个方向[35,36]。目前，GO 二维薄膜的成膜方法包括抽滤成膜法[37~40]、气液界面自组装法[41~46]、泡泡模板法[47]、层层（Layer by Layer）自组装法[48]、涂覆法[49]、Langmuir-Blodgett（LB）组装法[50]等多种方法。GO 薄膜在触摸屏、分子传感器、晶体管、电子和光电子器件等诸多领域得到了广泛应用[51]。下面将具体对 GO 薄膜的制备方法、性能和应用进行介绍。

4.2.1 抽滤成膜法

Ruoff 课题组[37]于 2007 年报道了利用抽滤成膜法制备薄膜的成果。该研究是以剥离的氧化石墨悬浮液为原料，通过真空抽滤法制备了柔性而且机械强度较高的 GO 纸，得到的 GO 纸的平均模量为 32GPa，最高值为 (42±2)GPa。真空抽滤法操作简单，可以获得相对均匀的薄膜材料，但是也存在抽滤过程缓慢、GO 纸不容易从滤膜上取下的问题。

GO 由于片层含有含氧官能团使其易于分散在水中。通过真空抽滤成膜，GO 片层又能够重新堆叠形成具有优异机械性能的纸状膜。研究发现，GO 膜在水中是非常稳定的；然而，这与 GO 片材在水中因水合作用带负电荷，电荷间静电排斥作用会克服范德华吸引力或氢键导致 GO 片材彼此分离的情况不符。Yeh 等[52]人报道了这一明显矛盾背后长期被忽视的原因。研究表明，洗涤较为干净的 GO 膜确实很容易在水中崩解，但如果 GO 膜被多价金属阳离子交联，则膜会变得稳定。这些金属阳离子污染物可能在 GO 合成和加工过程中被无意引入，其中最常见的引入情况是在用多孔阳极氧化铝（Anodic Aluminum Oxide，简称 AAO）过滤膜过滤时，由于氧化铝膜被腐蚀而释放的大量铝离子进入过滤液导致。Yeh 等[52]人对比了通过多孔 AAO 过滤器、硝酸纤维素滤膜或 Teflon 滤膜经真空过滤法制备的 GO 膜在水中的稳定性。该研究表明，通过多孔 AAO 过滤器得到的 GO 膜，经过 1 天浸泡后，依然很稳定。研究还发现，GO 薄膜的性质与分散液的 pH 值有很大的相关性。在高酸性（pH<4）、近中性（4<pH<8）和高碱性（pH>8）溶液中可以识别出三个腐蚀行为区域,这些溶液分别释放出 Al^{3+}、$Al(OH)_3$ 和 $Al(OH)_4^-$。由于 Al^{3+} 的交联作用,酸性 GO 分散液(pH=3)通过 AAO 抽滤制备的 GO 膜在水中具有高度的稳定性,而从近中性(pH=5.5)和碱性溶液中(pH=8.5)获得的薄膜虽然也含有 Al^{3+},但它们仍会在水中崩解,可能由于中性 $Al(OH)_3$ 和阴离子 $Al(OH)_4^-$ 不能交联带负电的 GO 片材。这一发现对于解释 GO 和其他层状膜的加工-结构-性质关系具有重要意义。

Nandy 等[53]人通过分段凝固技术揭示了真空辅助自组装制备的 GO 纸层状结构的形成机制。通过快速冷冻正在进行过滤的 GO 悬浮液,证实了在介质悬浮中的 GO 纳米片和最终的自支撑 GO 纸之间存在半有序聚集体(Semi-Ordered Aggregates,简称 SOA),以及在 GO 纸形成过程中 GO 纳米片之间存在复杂的相互作用。SOA 本身由扩散结构层以及更致密的表层两个不同的区域组成,其中扩散结构层是由 GO 片堆叠形成,而致密的表皮层会阻止过滤过程中的水流。作者为了描述随着时间变化 GO 纸的形成机制,使用凝固技术在过滤过程中的各个时间点拍摄了几个样本的图像。在施加真空抽滤之后,水开始快速地流过,在过滤器表面附近的纳米薄片是密闭的。图 4.7(a) 显示,在仅 2% 的水过滤出时,过滤器表面附近的纳米片排列成厚度约为 $10\mu m$ 的 SOA,平行于过滤器表面排列。5% 的水过滤出后,过滤速率减慢,图 4.7(b) 显示出通过 GO 层状前驱体或原始薄片聚集体的额外堆叠,SOA 持续生长至数十微米。尽管单个 GO 纳米片非常容易堆叠和起皱,但抽滤薄膜由于其厚度大,所以在机械强度上更硬,因此可以承受更大变形。多个初始薄膜最终聚集形成的薄膜构成了最终得到的 GO 纸中间大尺度结构。通过结构去除水而使初始层状压实与初始薄膜机械变形的阻力之间的竞争,导致 SOA 作为多孔网络的原始薄片持续生长。如图 4.7(b)、(c) 所示,过滤膜表面附近的 SOA 显示出了 60~70nm 厚的"皮肤"层,由数十个高度压实的纳米片组成,其外观类似于完全压实的 GO 纸。我们将这个松散有序的初始薄膜区域称为皮肤层,即"扩散结构"层。扩散结构层和表层一起构成 SOA。如图 4.7(d) 所示,当过滤完成 25% 时,SOA 达到数百微米高度。图 4.7(e)、(f) 中可以观察到,过滤膜表面附近的结构显示底部 GO 层的生长达到几微米。该扩散结构层持续至过滤程度为 75%,参照图 4.7(g),甚至过滤程度达到 90%[图 4.7(h)]。在悬浮液过滤完成 90% 后,底部表层的厚度为 $7\mu m$,大约为完全组装的 GO 纸厚度的三分之二。随后,空气/液体界面到达扩散结构层的顶部,此时,底部皮层上方的整个结构由互连的层状网络组成,如图 4.7(h)、(i) 所示,类似于在过滤的初始阶段 [图 4.7(a)、(b)],但厚度增加了三个数量级。由空气/液体界面施加的附加约束迫使初始层状网络压缩成完全成形的薄膜 [图 4.7(l)],厚度约为 25nm。作者认为这种阶段冷冻方法可扩展到其他薄膜结构的液相自组装;通过这种方法直接阐明材料形成机制,也可为改进和调整各种自组装和自支撑材料性能开辟出新途径。

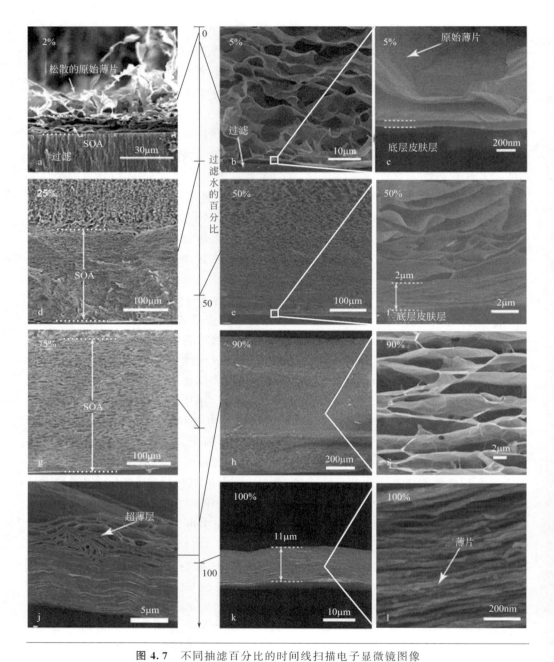

图 4.7 不同抽滤百分比的时间线扫描电子显微镜图像

图像百分比:(a) 2%、(b) 5%、(c) 5%、(d) 25%、(e) 50%、(f) 50%、(g) 75%、(h) 90%、(i) 90%、(j) 约 99% 和 (k)、(l) 100%(SOA:半聚集体)[53]

Chen 等[54]人通过热处理 GO 薄膜制备了"石墨"膜。这项工作讨论了在不同温度（300℃、1000℃或 2000℃）下热压的 GO 薄膜的压力辅助热分解，在 1000℃或 2000℃下热压的薄膜随后在 2750℃加热获得更高程度的石墨化。加热和加压的结合同时促进了 GO 膜的热还原和石墨化转变。在 300℃和 1000℃下热压的 GO 膜可以除去层间插入的水和含氧基团，并导致结构无序。由于没有氢键网络以及存在高度的结构缺陷，膜的机械强度很低，电导率随着热导率的降低而增加，石墨烯的导热系数对结构缺陷更敏感。在 2000℃下热压增加了有序石墨烯薄片堆叠的平均尺寸，并将石墨结构从不相称修复到更高程度的三维有序。在 2000℃热压、2750℃进一步石墨化的石墨烯薄膜的化学纯度高、均匀性及柔韧性更好，测得该薄膜的导电率约为 3.1×10^5 S/m，面内导热系数约为 1.2×10^3 W/m·K，机械性能优于报道的通过压延剥离石墨薄片制备的"石墨"膜。

4.2.2 Langmuir-Blodgett（LB）组装法

Langmuir-Blodgett 技术通常用于在基底上沉积纳米粒子和大分子来制备厚度可控的薄膜，也被用于制备石墨烯透明导电薄膜[55,56]。Cote 等[57]人发现通过 Langmuir-Blodgett 组装法，可以在不需要任何表面活性剂或稳定剂的情况下，利用 GO 层之间的静电排斥，在空气-水界面获得稳定的单层 GO。在制备单层 GO 的过程中，首先将 GO 分散在去离子水/甲醇（1:5）的混合溶液中，随着有机溶剂的挥发，GO 片层铺展到水表面上，形成单层。然后使用移动基底来改变单层的面积，从而有效地调节片间距离。当单层 GO 被限制在二维的空气-水界面时，能形成稳定的抗絮凝或凝结的分散。单层 GO 表现出可逆的等温压缩-膨胀循环。与分子和单层硬胶体颗粒相比，单层 GO 倾向于在边缘处折叠和起皱以抵抗塌陷成多层。单层 GO 可以转移到其他基底上，制备大面积的单层氧化石墨膜。GO 薄膜的密度可以从稀释、密堆积到单层互锁、单层过度填充之间连续调节。对于尺寸不匹配的单层，面与面的相互作用会导致不可逆的堆叠，形成双层。这种大面积具有连续可调密度的单层氧化石墨，可以通过化学还原转换为石墨烯薄膜，应用于电子领域，如透明导体薄膜。

2008 年，Li 等[58]人报道了用 LB 组装法逐层制备大尺寸石墨烯透明导电薄膜。具体的制备过程为：首先通过高温剥离可膨胀石墨制备单层石墨烯片材，然后，经过滤、超声等工艺获得均匀的石墨烯/DMF 悬浮液；再通过商业的 LB 槽制备石墨烯膜，通过高温煅烧除去膜上的残留有机物；最后，通过重复的 LB 法制备得到多层石墨烯膜。

Zheng 等[59]人通过 LB 自组装逐层沉积在衬底上制备了超大面积 GO（记作 UL-GO）透明导电膜，通过改变 LB 加工条件来控制 UL-GO 膜的密度和起皱程度，在沉积过程中形成了从单个片层 GO 膜到紧密堆积的扁平（UL-GO）的结构转变。研究表明，这种方法制备的薄膜经过还原后，薄膜具有 90% 的透明度，其薄层电阻约 500Ω/□，优于通过化学气相沉积在 Ni 基底上生长的石墨烯薄膜。这种制备透明导电 UL-GO 薄膜的技术操作简便，适用于规模化生产。

4.2.3 逐层组装法

逐层组装技术主要用于制造具有可调组成和结构的多层膜，这种方法的驱动力主要有静电作用、氢键和共价键。基于这几种驱动力，GO 可以与自身或者其他材料混合成膜。Byon 等[60]通过逐层组装法制备了碳纳米管和 RGO 的多层膜，由表面带负电荷的 RGO 和表面带正电荷的胺官能团化的多壁纳米碳管通过酰胺键交联组装而成，该薄膜可以应用于电化学微电容器。研究发现，这种薄膜组装的碳电极在酸性电解质（0.5M H_2SO_4）中具有约 160F/cm^3 的体积电容，主要归因于质子与碳上的表面含氧基团之间的氧化还原反应。掺入混合膜中的纳米碳管起间隔物的作用，可以避免由于 GO 基面之间的 π-π 相互作用引起的 RGO 的聚集。

Kumarasamy 等[61]人采用逐层组装法制备了具有三维交替层状纳米结构的金纳米粒子（AuNPs）和 RGO 的纳米杂化薄膜，这种薄膜可以用于电化学传感领域，通过金-硫醇相互作用捕获 DNA 并将其固定到具有自修复功能的 AuNP/RGO/AuNP/GCE（玻碳电极）薄膜的表面，然后将其作为标记的阻抗感测平台。这项研究可以应用于癌症诊断中的 DNA 杂交的无标记超灵敏检测。

4.2.4 界面诱导自组装

GO 纳米片的两亲特性有利于在液体/空气界面处形成定向组装，所以界面诱导自组装也是制备 GO 薄膜的一种常用的方法[41,62,63]。相比于其他制备 GO 膜的方法，这种方法更加简单高效。这种方法的原理是：当加热 GO 水溶胶时，由于布朗运动，GO 纳米片在悬浮液中的运动加剧；GO 水溶胶的液面随着加热过程的进行而逐渐下降，GO 纳米片随着水分子从水溶胶中的逸出而迁移到液体-空气界面，进而聚集形成多层结构。Chen 等[41]人报道了通过蒸发 GO 悬浮液在液体/空气界面处较快速地制备出无支撑 GO 膜的研究结果。在实验中，首先将浓度为 2mg/mL 的 GO 水性悬浮液超声处理 30 分钟，然后以 5000r/min 的速度高速离心 20 分钟除去杂质，将 GO 的水溶胶加热至 353K，在液-气界面处形成光滑且凝聚的 GO 薄膜。通过进一步的干燥和还原处理，可以制备厚度为 0.5~20μm 的柔性、半透明和自立式石墨烯纸（如图 4.8 所示），纸的厚度可以通过控制水溶胶的蒸发时间来调节。例如，加热 20 分钟和 40 分钟后分别获得厚度为 5μm 和 10μm 的 GO 纸。柔性 GO 纸的尺寸仅受液/气界面面积的限制，因此可以通过使用大尺寸反应容器获得大面积纸张。受到水的表面张力影响，该方法难以制备特别薄的无支撑 GO 薄膜。这种宏观膜由单个 GO 片通过逐层堆叠而成，显示出优异的机械和光学性能。研究发现，这种膜的弹性模量为 13.8GPa，拉伸强度为 75.9MPa。

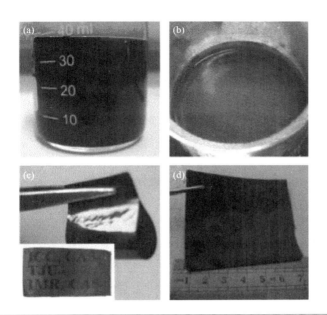

图 4.8 GO 薄膜的制备流程[41]

Shen 等[64] 人通过在温和加热条件（50~60℃）下直接蒸发 GO 悬浮液来制造 GO 薄膜，并且通过石墨化 GO 薄膜制备出超薄石墨烯薄膜。研究表明，得到的厚度仅 8.4μm 的石墨烯薄膜不仅具有优异的电磁屏蔽性能（约 20dB）和很高的面内热导率（约 1100W/m·K），而且还表现出优异的机械性能。图 4.9(a) 为 GO 薄膜的制造过程示意图：将 GO 悬浮液倒入平面容器中，在 50~60℃下加热将水蒸发，从而形成柔软的深棕色 GO 纸。通过控制 GO 悬浮液的体积和浓度，可以有效地控制纸张的厚度。图 4.9(b) 显示了简单地通过使用这种尺寸的模具制成的尺寸约为 400cm² 的 GO 纸的照片。GO 纸张足够柔软，可以卷起和折叠，如图 4.9(c) 所示。扫描电子显微镜观察结果表明，GO 纸具有平滑的几条薄薄的褶皱，如图 4.9(d) 所示。从图 4.9(e)、(f) 中可以看到，多层结构布满整个截面，和真空抽滤得到的薄膜类似。图 4.9(d) 显示了 GF-2000 样品的 SEM 图像，表明其表面光滑且存在褶皱。图 4.9(e)、(f) 显示了 GF-2000 样品的横截面 SEM 图像，说明多层堆叠的石墨烯纳米片层具有高度取向结构。

4.2.5　旋转涂膜法

旋转涂膜法是一种广泛使用的方法，可以很好地控制所制备薄膜的形态和微观结构，具有较好的重复性。Becerril 等[65] 人通过在石英上旋涂 GO 悬浮液制备了 GO 薄膜，再经过石墨化处理，制得了薄层电阻低至 $10^2 \sim 10^3$ Ω/□ 的薄膜，其中

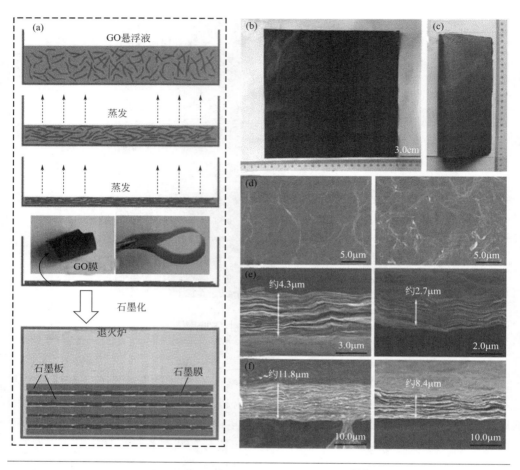

图 4.9 GO 膜的制备工艺及 SEM 表征[64]

550nm 光线的透过率为 80%，可以应用于透明电极。Beese 等[66] 人通过旋转涂膜法制备了 GO 和 PMMA 的复合膜，研究发现，这种多层 GO-PMMA 薄膜与纯 GO 薄膜相比机械性能更好，其刚度和强度增加高达 100%。固体含量、溶液黏度和旋涂条件等因素会影响薄膜的最终厚度。旋转涂膜法的缺点是不适于 GO 薄膜的大面积生产。

4.2.6 其他方法

除以上 GO 膜的制备方法外，泡泡模板法[47]、涂层棒法[67] 等二维组装方法也被用于制备 GO 二维结构。Chen 等[47] 人通过干燥由刚性框架支撑的液体制备了厘米级的 GO 薄膜，将该薄膜还原后获得薄的 RGO 薄膜。经测试，该薄膜的最大透明度为 75.8%，最小薄层电阻为 920Ω/□。Wang 等[67] 人通过杆式涂布技术结合室温还原，直接在柔性基底上大规模制备 RGO 薄膜（如图 4.10 所示）。所制备的薄膜显示出优异的均匀性、良好的透明度和导电性以及触摸屏的高度灵活性。

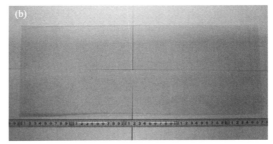
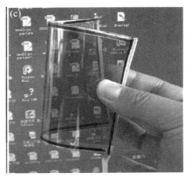

图 4.10 杆式涂布法制备的 GO 薄膜[67]

目前研究表明，在大部分的应用研究中需要将 GO 薄膜还原为 RGO 膜，而经过还原后得到 RGO 膜可以用作透明导电薄膜、分离膜、传感器、防腐涂层以及超级电容器、锂离子电池等能源领域[68~71]。后续章节将会对这些领域的应用做进一步的详细介绍。

4.3 氧化石墨烯三维宏观体结构组装及应用

在很多应用场景中，需要制备三维宏观材料，因而三维 GO 宏观体材料日益受到关注，尤其是三维 GO 泡沫或块体材料已经成为目前的研究热点。实际上，从低维单元到三维材料，通常得到的 GO 宏观材料是相互联系的统一体，而且可以根据应用领域来调节制备参数，能够实现几种宏观形态的转换[72~74]。目前，三维 GO 宏观体材料的制备方法主要通过利用各种非平衡态过程中的自组装，或者在模板/

官能团帮助/引导下的组装实现,包括水热法、模板法、交联法、化学还原自组装法、电化学沉积法、定向流动自组装法等。以下将主要简述通过这些制备方法得到的 GO 三维宏观体结构的相关研究结果。

4.3.1 水热法

GO 含有疏水基底面和亲水性含氧基团(如羟基、环氧基、羰基和羧基),这有助于 GO 通过共价键和非共价键进行功能化[75],这种两亲性也为 GO 实现三维自组装提供了可能。凝胶是三维 GO 最常见的一种结构形态,水热法是一种常用的还原 GO 的方法,也是生产石墨烯水凝胶中的常用方法。例如,Xu 等[76] 人通过一步水热法合成了自组装石墨烯水凝胶 [图 4.11(a)、(b)],将浓度不低于 2mg/mL 的 GO 分散液在 180℃下水热处理 12h,经过冷冻干燥后,所制备的石墨烯水凝胶的电导率可以达到 (5 ± 0.2)mS/cm,模量高达 450~490kPa,比电容可以达到 (160 ± 5)F/g。形成的石墨烯水凝胶为 3D 多孔网络结构 [图 4.11(c)]。研究者提出如下形成机理:在水热还原过程中,GO 片层上的含氧官能团被部分去除,还原后 GO 的疏水性提高,并且还原后 GO 片材的 π-π 共轭结构更紧密,这增强了片层之间的相互作用,又保持了残余官能团,以便捕获充足的水到石墨烯网络中。

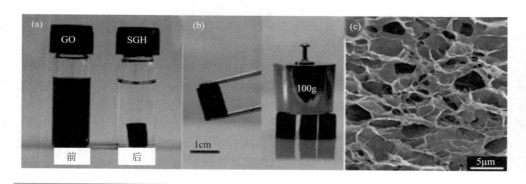

图 4.11 一步水热法合成自组装石墨烯水凝胶及其多孔网络结构[76]

Zhang 等[77] 人进一步用肼或氢碘酸化学还原了石墨烯水凝胶,见图 4.12,可以进一步提高其电导率。经再次还原后,这种网络结构水凝胶的电导率达到 1.3~3.2S/m。用肼还原的石墨烯水凝胶超级电容器在 1A/g 下表现出 220F/g 的比电容;当放电电流密度增加到 100A/g 时,该电容可以保持 74%。这种高功率密度和长循环寿命的超级电容器可应用于高速充电/放电储能等领域。

此外,大量的研究集中在金属盐[78]、金属氧化物纳米粒子[79,80]、聚吡咯纳米

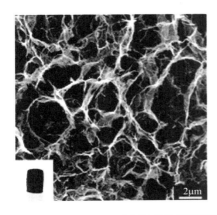
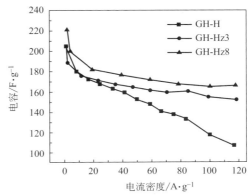

图 4.12 化学还原后的石墨烯水凝胶微观形貌及其超级电容器充/放电性能测试[77]

粒子[81]或其他元素存在的条件下，通过一步水热自组装法制备 GO 三维材料，用于高效的电催化剂、锂离子电池阳极和药物载体。Jiang 等[82]人通过水热法将 RGO 微片与二价离子和水分子结合制备了凝胶状 3D RGO 结构。Wu 等[83]人通过水热反应制备了以 GO 和氨三氟化硼为前驱体的 3D 氮和硼共掺杂石墨烯气凝胶，并利用这种气凝胶制备了高性能全固态超级电容器。研究表明，该超级电容器的比电容、能量密度和功率密度均较高。Huang 等[84]人报道了两步自组装制备三维石墨烯/SnO_2 的方法。具体组装过程为：首先将 GO 溶液与 SnO_2 纳米颗粒混合均匀；第二步，负载 SnO_2 纳米颗粒的 GO 经过水热自组装成三维石墨烯/SnO_2 复合材料。合成的三维石墨烯/SnO_2 复合材料具有高比表面积及低密度。而且，这种特殊的结构在锂离子电池循环过程中能够容纳 SnO_2 粒子的体积膨胀，同时为电子运输和电解液通路提供了多维度通道，可使锂离子从电解液快速扩散到电极。因而三维石墨烯/SnO_2 复合材料在 100mA/g 的电流密度下循环 70 次后依然具有 830mA·h/g 的高容量。通过水热法获得的高表面积 3D 石墨烯结构也在超级电容器领域展现了应用可能性[85,86]。

4.3.2 溶剂热法

类似地，利用溶剂热还原也能在还原 GO 悬浮液的同时对其进行组装，区别是溶剂热法通常用电化学稳定性更高的有机电解质取代水性介质。Liu 等[87]人通过在碳酸亚丙酯中还原 GO 分散液制备了一种三维大孔结构石墨烯有机凝胶，其孔径在亚微米至几微米范围内，比表面积为 $146m^2/g$，电导率为 2S/m。溶剂热过程也可以将各种掺杂元素加入三维石墨烯结构中[87,88]。例如，Liu 等[87]人通过一步法溶剂热处理 $SnCl_4$ 和石墨烯氧化物的混合物，随后进行冷冻干燥制备了负载 SnO_2 纳米粒子的 3D 石墨烯气凝胶。这种石墨烯气凝胶具有互通大孔网络结构，其中均匀分布着尺寸为 5~10nm 的 SnO_2 纳米粒子，可以用于气体检测。室温时，这种复合气凝胶对 NO_2 具有优异的响应和选择性。优异的灵敏度被归因于石墨烯气凝胶的大比表面积，从而使目标气体分子与表面活性位点之间形成更多有效的表面相互作用。

Ma 等[88]人开发出一种简便且可控的方法来合成基于 3D 石墨烯的块体状材料。通过一步溶解热还原法，组装成了三维多孔氧化钴/石墨烯片（CoO/GS）复合材料。在复合材料中，CoO 颗粒周围的自组装三维石墨烯网络不仅可以增强基体的导电性，还可以缓冲 CoO 在充放电过程中的体积变化，并能够阻止 CoO 颗粒在充放电循环过程中的聚集，同时均匀锚固到 CoO/GS 复合材料框架中的 CoO 粒子也可以充当间隔物，有效地避免石墨烯片层的重叠。与纯 CoO 相比，3D CoO/GS 复合材料作为锂离子电池阳极显示出优异的电化学性能，包括循环稳定性和倍率性能；经过 50 次循环后，在 1600mA·h/g、4800mA·h/g 和 6400mA·h/g 的电流密度下仍可分别保持约 706mA·h/g、503mA·h/g 和 405mA·h/g 的稳定可逆容量，而未复合 3D 石墨烯的 CoO 在 1600mA·h/g 的电流密度下循环 50 次后，其可逆容量仅为 211mA·h/g。

4.3.3 化学还原中的组装

如第 2 章所述，化学还原 GO 不需要在惰性环境、高温和高压的条件下进行，而可以使用一些比较温和的还原剂（如肼，维生素 C，抗坏血酸钠[89~91]，HI，Na_2S 和 $NaHSO_3$[92]，L-抗坏血酸[93]，抗坏血酸钠[89]，L-谷胱甘肽[94] 以及草酸和碘化钠的组合[95]）来恢复石墨烯的 sp^2 结构。通过还原剂还原后的石墨烯水凝胶通过冷冻干燥处理可以得到石墨烯气凝胶。Zhang 等[95]人使用草酸和碘化钠（NaI）作为还原剂还原 GO，通过简便的原位还原-组装方法制备了 3D 石墨烯凝胶，这种方法被认为是一种成本低而且环保的制备方法。在这项研究中，甲酸和 NaI 的组合可以将各种浓度（0.1~4.5mg/mL）的 GO 悬浮液制备成 3D 石墨烯凝胶。而且制备的 3D 石墨烯凝胶表现出低密度和良好的导电性能，4.5mg/mL 的三维石墨烯凝胶的电导率为 24.8S/m。该研究还将聚二甲基硅氧烷渗入到 3D 石墨烯导电网络，制备了响应性导电聚合物，可以用来区分不同极性的有机溶剂。

4.3.4 三维组装中的交联方法

在交联法中，采用聚合物[96~98]、离子键[99] 和 DNA[100] 等交联剂，通过调整交联剂的结构，可以将 GO 凝胶浇铸成所需的尺寸或形状，并且力图实现孔径可控。例如，Xu 等[100]人用 DNA 组装 GO 片来制造 3D 多功能石墨烯基水凝胶，研究发现，该水凝胶具有高的机械强度，环境稳定性和染料加载能力，并显示出自愈合特性。在水凝胶中，GO 和 DNA 生物大分子链通过 π-π 键和疏水作用缠结在一起构成三维复合材料，如图 4.13 所示。

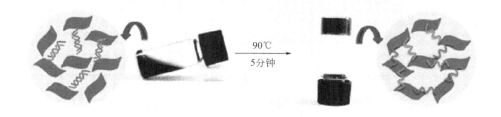

图 4.13 DNA 组装 GO 片制备多功能石墨烯基水凝胶示意[100]

Wan 等[101]人利用 GO 片表面上的环氧基团与聚氧丙烯二元胺（D400）上的氨基共价交联，制备出通过共价键连接的三维石墨烯/D400 复合材料，这种方法反应时间仅为 60 秒。通过调节 D400 与 GO 的比例，可以产生 3 种不同的 GO 组装结构。而且由此制备的三维石墨烯/D400 复合材料可以被塑造成各种形状。制造的 GOM 对 L929 细胞系无细胞毒性，可以应用于生物医学领域。

Wang 等[102]人的研究表明，静电作用也可以驱动 GO 片与聚合物组装。GO 片层带负电，因此可以和带正电的聚合物通过静电作用组装。研究者通过静电作用制备了 GO 聚乙烯亚胺（Polyetherimide，简称 PEI）凝胶，经组装的 GO/PEI 水凝胶，无须再进行任何的二次化学交联。在组装过程中，调整 GO 和 PEI 的比例，可以制备一系列的水凝胶，尽管缺乏化学交联，但是在过量的水或 NaCl 溶液中时，GO/PEI 水凝胶是稳定的。水凝胶的不可逆机械压缩可使其储能模量增加超过 3 个数量级，能够用于检测水凝胶对预先加载的压缩产生的机械刺激响应。水凝胶这一系列的机械适应性能够应用于传感器、能量耗散装置和其他领域中。

Wang 等[103]人将 GO 添加到 PVA 中来调整复合材料合成过程中的形态以及孔结构。研究发现，PVA 和 GO 之间强的氢键作用，能够提高 PVA 复合材料的热稳定性和吸水能力。利用 Brunauer-Emmett-Teller（BET）吸附方法计算得到，当 GO 的添加量分别为 20mg/mL 和 50mg/mL 时，可使 PVA 的比表面积由 $115m^2/g$ 分别增加到 $134m^2/g$ 和 $150m^2/g$。这种 GO-PVA 复合材料可以用于除油、催化等领域。

Wang 等[104]人通过涂覆在多孔和自支撑三维石墨烯泡沫上的聚苯胺（Polyaniline，简称 PANI）的氧化聚合，开发了具有高导电性和机械性能的轻质、易加工环氧树脂复合材料。在制备过程中，首先利用 PVA 和 GO 之间强的氢键作用制备均匀稳定的 GO/PVA 分散体，然后通过冷冻干燥、高温热还原除去分散体中的 PVA，得到三维石墨烯泡沫。再通过原位化学聚合方法制备 PANI 改性的三维石墨烯泡沫（3D GF-PANI）。将制备好的 3D GF-PANI 放入模具中，然后用真空泵将含固化剂的环氧树脂抽入模具中。环氧树脂在一定温度下固化，得到环氧树脂/3D

GF 复合材料。研究发现，对于含有 0.39%（质量）3D GFs-PANI 的环氧树脂复合材料，导电率比纯环氧树脂提高了近 12 个数量级，达到 0.036S/cm。此外，机械性能测试表明，环氧/3D GF-PANI 复合材料的抗压强度分别比环氧树脂/3D G 复合材料和固态纯环氧树脂提高 72.6% 和 25%。3D GFs 在环氧树脂内部相互连通形成导电网络，石墨烯片与 PANI 之间的强 π-π 作用以及环氧和 PANI 链之间的化学键合提高了 3D GFs 和环氧树脂之间的界面结合，成为提高复合材料导电性和力学性能的主要原因。

4.3.5 模板法

模板法是将胶束[105,106]、乳胶[107~110]、气泡[111] 和高分子[112] 等模板结构引入 GO 悬浮液中，来控制所制备的三维材料的形貌和内部结构。材料成型后，再通过过滤、还原或冷冻等技术将 GO 微片固定在模板的界面处，最后将模板去除得到三维 GO 宏观体结构。Huang 等[106] 人通过乳液软模板法合成用于锂-氧电池和吸附领域的多孔石墨烯泡沫材料。该研究中，作者首先分别采用 3,3′,5,5′-四甲基联苯胺和正十六烷与 HCl 溶液混合搅拌、超声后得到混浊悬浮液，再将这两种悬浮液分别与石墨烯氧化物溶液、水混合，通过真空过滤收集沉淀物，将得到的沉淀物首先在 350℃ 的氩气中煅烧 5 小时，再在氩气气氛下在 900℃ 下再加热 5 小时后得到多孔石墨烯泡沫材料。其中通过 3,3′,5,5′-四甲基联苯胺与 GO 制备得到的多孔石墨烯泡沫主要用于锂氧电池中的阴极材料，通过正十六烷与 GO 制备得到的多孔石墨烯泡沫主要用于吸附，表现出高效率的吸油能力。当用作锂-氧电池中电极材料时，多孔石墨烯泡沫显示出良好的催化活性。

Menzel 等[110] 人通过乳液模板法制备了具有限定形状的热 RGO 气凝胶（密度为 10mg/cm^3）。这种气凝胶的制备过程为：将 GO 与有机添加剂（聚乙烯醇：蔗糖=1：1）混合均匀后制成纳米碳悬浮液；将纳米碳悬浮液与疏水相甲苯进行乳化得到乳液，将乳液浇铸到圆柱形聚四氟乙烯模具中，通过冷冻干燥获得块状纳米碳材料；然后在管式炉内（Ar，1000℃）热还原制备成 RGO 气凝胶。这种超轻纳米碳气凝胶在较低电压（约 1V）和电源输入（约 2.5W/cm^3）条件下可重复加热至 200℃（图 4.14），具有快速的加热和冷却速率（高达 10K/s），可以被用于快速和可重复温度循环的气体加热器领域。

Li 等[108] 人将己烷溶液滴入到 GO 分散体中，随后进行水热处理。在水热还原过程中，石墨烯微片围绕己烷液滴组装成网络结构，再经过冷冻干燥，得到弹性和导电性都很好的石墨烯块状物。Su 等[112] 人将 PMMA（聚甲基丙烯酸甲酯）纳米颗粒作为模板引入到 GO 溶液中，抽滤后获得夹杂 PMMA 纳米颗粒的薄膜，再经 800℃ 高温热处理除去高分子，并同时对氧化石墨烯还原，最后获得内部有球形结构的三维石墨烯结构。

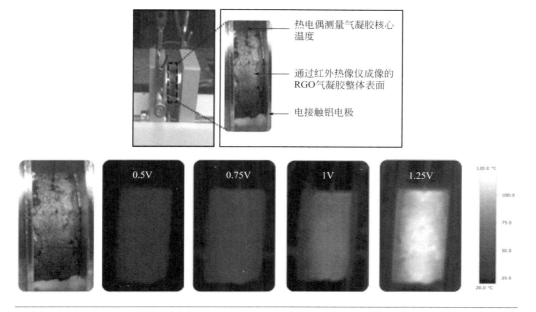

图 4.14 低电压下 RGO 气凝胶的红外热相图[110]

4.3.6 其他方法

除了以上几种利用 GO 悬浮液制备 GO 三维宏观体材料的方法，还有通过 GO 薄膜制备的 GO 三维宏观体材料的方法，如发酵法、激光照射法、流延成型法等。

Niu 等[113]受面团制作过程的启发，采用发酵法将多层 GO 结构转化为多孔和交联网络的导电 RGO 泡沫。由于还原过程生成了气体，对石墨烯产生了明显的膨胀和造孔效应，所以制得的材料为泡沫结构，而且所获得的 RGO 泡沫是柔性的，其拉伸强度约为 3.2MPa，初始和最后拉伸阶段的杨氏模量分别约为 7MPa 和 40MPa。该材料的拉伸应变可高达约 10%，这比 GO 纸（0.6%）的拉伸应变大得多。这种 RGO 泡沫可以用作超级电容器和柔性电极材料。

El-Kady 等[114,115]人通过 Light Scribe DVD 光驱在柔性衬底上将氧化石墨膜直接激光还原成石墨烯薄膜。由此得到的石墨烯薄膜表现出优异的导电性（1738S/m）和较高的比表面积（1520m²/g），同时还呈现出网络结构。Teoh 等[116]人使用聚焦激光束将 GO 转化为 RGO，并在加热的基底上使用 GO 旋涂微图案制备成 3D 层状石墨烯结构，类似层层自组装过程。Korkut 等[117]人通过流延法制备了长 0.1m、宽 15cm 的功能化石墨烯片（FSP）/聚合物/表面活性剂复合片材。制备过程如图 4.15 所示，将复合片材从石英基板上取下，再通过加热除去大部分的聚合物和表面活性剂，形成独立和连续的功能化三维石墨烯体结构。这种三维石墨烯结构厚度均匀，微观结构无缺陷，并且其比表面积较高（>400m²/g），表观密度介于 0.15～0.51g/cm³ 之间，电导率高达 24kS/m，拉伸强度超过 10MPa，其结构和性能还随着 FSP/聚合物/表面活性剂的浓度而变化。相对于其他方法而言，流延法更加简单易行，易于实现工业化生产。

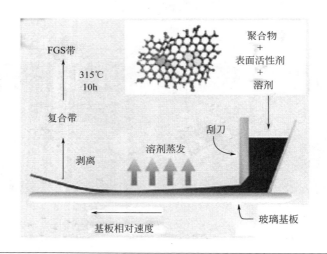

图 4.15 流延法制备三维石墨烯材料[117]

三维 GO 宏观结构与一维 GO 纤维、二维 GO 薄膜结构在高导电性方面具有共同特性，可以用于超级电容器、电化学传感器、吸附材料、催化剂载体等[118～120]。

4.4 小结

本章节对于以 GO 为主体材料组装成的一维纤维、二维薄膜以及三维宏观材料的制备、性质和应用进行了介绍。大量研究已经证明，通过自组装过程的调控，可以实现氧化石墨烯材料结构的控制，并已开发出一系列具有独特结构和优异性能的碳基功能材料，这些材料在众多领域表现出良好的应用潜力。相信随着未来研究的深入，自组装氧化石墨烯材料可以在不同领域实现大规模的应用。

—— 参考文献 ——

[1] Yu Dingshan, Yuan Zhongke, Xiao Xiaofen, et al. Self-Assembled Graphene Nanostructures and Their Applications [J]. Functional Organic Hybrid Nanostructured Materials: Fabrication, Properties, Applications, **2018**.

[2] Li Zheng, Liu Zheng, Sun Haiyan, et al. Superstructured assembly of nanocarbons: fullerenes, nanotubes, and graphene [J]. Chemical reviews, **2015**, 115 (15): 7046-7117.

[3] Nuray Uçar, Ölmez Mervin, Kayaoğlu Burçak Karagüzel, et al. Structural properties of graphene oxide fibers: from graphene oxide dispersion until continuous graphene oxide fiber [J]. The Journal of The Textile Institute, **2018**: 1-11.

[4] Liu Yingjun, Hui Liang, Zhen Xu, et al. Superconducting Continuous Graphene Fibers via Calcium-Intercalation [J]. Acs Nano, **2017**, 11 (4): 4301-4306.

[5] Meng Fancheng, Lu Weibang, Li Qingwen, et al. Graphene-Based Fibers: A Review [J]. Advanced Materials, **2015**, 27 (35): 5113-5131.

[6] Lunev V. D., Machalaba N. N. Evaluation of the requirements for the rheological characteristics of the polymer solution for production of high-strength polyaramid fibres [J]. Fibre Chemistry, **2009**, 41 (5): 294-297.

[7] Zakharchenko Konstantin, Fasolino A, H Los J, et al. Melting of graphene: from two to one dimension [J]. Journal of Physics Condensed Matter An Institute of Physics Journal, **2011**, 23 (20): 202202.

[8] Xu Zhen, Sun Haiyan, Zhao Xiaoli, et al. Ultrastrong fibers assembled from giant graphene oxide sheets [J]. Advanced Materials, **2013**, 25 (2): 188-193.

[9] Jalili Rouhollah, Aboutalebi Seyed Hamed, Esrafilzadeh Dorna, et al. Graphene Oxide: Scalable One-Step Wet-Spinning of Graphene Fibers and Yarns from Liquid Crystalline Dispersions of Graphene Oxide: Towards Multifunctional Textiles (Adv. Funct. Mater. 43/2013) [J]. Advanced Functional Materials, **2013**, 23 (43): 5344.

[10] Behabtu Natnael, C Young Colin, Tsentalovich Dmitri, et al. Strong, Light, Multifunctional Fibers of Carbon Nanotubes with Ultrahigh Conductivity [J]. Science, **2013**, 339 (6116): 182-186.

[11] Balandin Alexander, Ghosh Suchismita, Bao Wenzhong, et al. Superior thermal conductivity of single-layer graphene [J]. Nano Letters, **2008**, 8 (3): 902-907.

[12] Liang Qizhen, Yao Xuxia, Wang Wei, et al. A three-dimensional vertically aligned functionalized multilayer graphene architecture: an approach for graphene-based thermal interfacial materials [J]. Acs Nano, **2011**, 5 (3): 2392-2401.

[13] Eom Wonsik, Park Hun, Noh Sung Hyun, et al. Strengthening and Stiffening Graphene Oxide Fiber with Trivalent Metal Ion Binders [J]. Particle Particle Systems Characterization, **2017**, 34 (9): 1600401.

[14] Zhang Dong, Peng Li, Shi Naien, et al. Self-assembled high-performance graphene oxide fibers using ionic liquid as coagulating agent [J]. Journal of Materials Science, **2017**, 52 (13): 7698-7708.

[15] Cao Jun, Zhang Yong, Men Chuanling, et al. Programmable writing of graphene oxide/reduced graphene oxide fibers for sensible networks with in situ welded junctions [J]. Acs Nano, **2014**, 8 (5): 4325-4333.

[16] Li Dan, B Müller Marc, Gilje Scott, et al. Processable aqueous dispersions of graphene nanosheets [J]. Nature Nanotechnology, **2008**, 3 (2): 101-105.

[17] Xiang Changsheng, Behabtu Natnael, Liu Yaodong, et al. Graphene nanoribbons as an advanced precursor for making carbon fiber [J]. Acs Nano, **2013**, 7 (2): 1628-1637.

[18] Dong Zelin, Changcheng Jiang, Cheng Huhu, et al. Facile fabrication of light, flexible and multifunctional graphene fibers [J]. Advanced Materials, **2012**, 24 (14): 1856-1861.

[19] Yu Dingshan, Goh Kunli, Wang Hong, et al. Scalable synthesis of hierarchically structured carbon nanotube-graphene fibres for capacitive energy storage [J]. Nature Nanotechnology, **2014**, 9 (7): 555.

[20] Li Jihao, Li Jingye, Li Linfan, et al. Flexible graphene fibers prepared by chemical reduction-induced

self-assembly [J]. Journal of Materials Chemistry A,**2014**,2 (18): 6359-6362.

[21] Chen Qing, Meng Yuning, Hu Chuangang, et al. MnO_2-modified hierarchical graphene fiber electrochemical supercapacitor [J]. Journal of Power Sources,**2014**,247 (3): 32-39.

[22] Acik Muge, Lee Geunsik, Mattevi Cecilia, et al. The Role of Oxygen during Thermal Reduction of Graphene Oxide Studied by Infrared Absorption Spectroscopy [J]. Journal of Physical Chemistry C,**2015**,115 (40): 19761-19781.

[23] Hu Chuan-Gang, Zhao Yang, Cheng Huhu, et al. Graphene microtubings: controlled fabrication and site-specific functionalization [J]. Nano Letters,**2012**,12 (11): 5879-5884.

[24] Fan Jing, Dong Zelin, Qi Meiling, et al. Monolithic graphene fibers for solid-phase microextraction [J]. Journal of Chromatography A,**2013**,1320 (20): 27-32.

[25] Seyedin Shayan, Romano Mark S., Minett Andrew I., et al. Towards the Knittability of Graphene Oxide Fibres [J]. Scientific Reports,**2015**,5: 14946.

[26] Yun Jang Eui, Gonzalez Javier, Choi Ajeong, et al. Fibers of reduced graphene oxide nanoribbons [J]. Nanotechnology,**2012**,23 (23): 235601.

[27] Rodolfo Cruz Silva, Aaron Morelos Gomez, Hyung-Ick Kim, et al. Super-stretchable graphene oxide macroscopic fibers with outstanding knotability fabricated by dry film scrolling [J]. Acs Nano,**2014**,8 (6): 5959-5967.

[28] Javier Carretero González, Elizabeth Castillo Martinez, Marcio Dias Lima, et al. Oriented graphene nanoribbon yarn and sheet from aligned multi-walled carbon nanotube sheets [J]. Advanced Materials,**2012**,24 (42): 5695-5701.

[29] Wang Siliang, Liu Nishuang, Su Jun, et al. Highly Stretchable and Self-Healable Supercapacitor with Reduced Graphene Oxide Based Fiber Springs [J]. Acs Nano,**2017**,11 (2): 2066-2074.

[30] Liang Kou, Chao Gao. Bioinspired design and macroscopic assembly of poly (vinyl alcohol) -coated graphene into kilometers-long fibers [J]. Nanoscale,**2013**,5 (10): 4370-4378.

[31] Aboutalebi Seyed Hamed, Jalili Rouhollah, Esrafilzadeh Dorna, et al. High-performance multifunctional graphene yarns: toward wearable all-carbon energy storage textiles [J]. Acs Nano,**2014**,8 (3): 2456-66.

[32] Fan Tianju, Zhao Chunyan, Xiao Zhuangqing, et al. Fabricating of high-performance functional graphene fibers for micro-capacitive energy storage [J]. Scientific Reports,**2016**,6: 29534.

[33] Meng Jie, Nie Wenqi, Zhang Kun, et al. Enhancing Electrochemical Performance of Graphene Fiber-Based Supercapacitors by Plasma Treatment [J]. Acs Applied Materials nterfaces,**2018**,10 (16): 13652-13659.

[34] Cheng Huhu, Hu Yue, Zhao Fei, et al. Moisture-activated torsional graphene-fiber motor [J]. Advanced Materials,**2014**,26 (18): 2909-2913.

[35] Sun Meng, Li Jinghong. Graphene oxide membranes: Functional structures, preparation and environmental applications [J]. Nano Today,**2018**,20: 121-137.

[36] Jilani Asim, Othman Mohd Hafiz Dzarfan, Ansari Mohammad Omaish, et al. A simple route to layer-by-layer assembled few layered graphene oxide nanosheets: Optical, dielectric and antibacterial aspects [J]. Journal of Molecular Liquids,**2018**,253: 284-296.

[37] Dikin Dmitriy A, Sasha Stankovich, Zimney Eric J, et al. Preparation and characterization of graphene oxide paper [J]. Nature,**2015**,448 (7152): 457-460.

[38] Chen Haiqun, Müller Marc B., Gilmore Kerry J., et al. Mechanically Strong, Electrically Conductive, and Biocompatible Graphene Paper [J]. Advanced Materials,**2010**,20 (18): 3557-3561.

[39] Tang Longhua, Wang Ying, Li Yueming, et al. Preparation, Structure, and Electrochemical Properties of Reduced Graphene Sheet Films [J]. Advanced Functional Materials, 2010, 19 (17): 2782-2789.

[40] Cong Huai-Ping, Ren Xiaochen, Wang Ping, et al. Flexible graphene-polyaniline composite paper for high-performance supercapacitor [J]. Energy Environmental Science, 2013, 6 (4): 1185-1191.

[41] Chen Cheng-Meng, Yang Q.-H, Yang Y, et al. Self-assembled free-standing graphite oxide membrane [J]. Advanced materials, 2009, 21 (29): 3007-3011.

[42] Chen Cheng Meng, Huang Jia Qi, Zhang Qiang, et al. Annealing a graphene oxide film to produce a free standing high conductive graphene film [J]. Carbon, 2012, 50 (2): 659-667.

[43] Shao Jiao-Jing, Lv Wei, Guo Quangui, et al. Hybridization of graphene oxide and carbon nanotubes at the liquid/air interface [J]. Chemical Communications, 2012, 48 (31): 3706-3708.

[44] Wu S. D., Lv Wei, Xu Jia, et al. A graphene/poly (vinyl alcohol) hybrid membrane self-assembled at the liquid/air interface: Enhanced mechanical performance and promising saturable absorber [J]. Journal of Materials Chemistry, 2012, 22 (33): 17204-17209.

[45] Lv Wei, Xia Zhang Xun, Wu Si Da, et al. Conductive graphene-based macroscopic membrane self-assembled at a liquid-air interface [J]. Journal of Materials Chemistry, 2011, 21 (10): 3359-3364.

[46] Shao Jiao-Jing, Lv Wei, and Yang Quan-Hong. Self-assembly of graphene oxide at interfaces [J]. Advanced Materials, 2015, 26 (32): 5586-5612.

[47] Chen Wufeng, Yan Lifeng. Centimeter-sized dried foam films of graphene: preparation, mechanical and electronic properties [J]. Advanced Materials, 2012, 24 (46): 6229-6233.

[48] Kotov Nicholas A., Fendler Janos H. Ultrathin graphite oxide-polyelectrolyte composites prepared by self-assembly: Transition between conductive and non-conductive states [J]. Advanced Materials, 1996, 8 (8): 637-641.

[49] Pham Viet Hung, Cuong Tran Viet, Hur Seung Hyun, et al. Fast and simple fabrication of a large transparent chemically-converted graphene film by spray-coating [J]. Carbon, 2010, 48 (7): 1945-1951.

[50] Kim Franklin, J Cote Laura, Huang Jiaxing. Graphene oxide: surface activity and two-dimensional assembly [J]. Advanced Materials, 2010, 22 (17): 1954-1958.

[51] Gao Hongcai, Duan Hongwei 2D and 3D graphene materials: Preparation and bioelectrochemical applications [J]. Biosensors Bioelectronics, 2015, 65: 404-419.

[52] Yeh Che-Ning, Raidongia Kalyan, Shao Jiaojing, et al. On the origin of the stability of graphene oxide membranes in water [J]. Nature Chemistry, 2014, 7 (2): 166-170.

[53] Nandy K, J. Palmeri M, M. Burke C, et al. Stop Motion Animation Reveals Formation Mechanism of Hierarchical Structure in Graphene Oxide Papers [J]. Advanced Materials Interfaces, 2016, 3 (6): 1500666.

[54] Chen Xianjue, Deng Xiaomei, Na Yeon Kim, et al. Graphitization of graphene oxide films under pressure [J]. Carbon, 2018, 132: 294-303.

[55] Jin Young Park, Advincula Rigoberto C. Nanostructuring polymers, colloids, and nanomaterials at the air-water interface through Langmuir and Langmuir-Blodgett techniques [J]. Soft Matter, 2011, 7 (21): 9829-9843.

[56] J Cote Laura, Cruz-Silva Rodolfo, Huang Jiaxing. Flash reduction and patterning of graphite oxide and its polymer composite [J]. Journal of the American Chemical Society, 2009, 131 (31): 11027-11032.

[57] Cote Laura J, Kim Franklin, Huang Jiaxing. Langmuir-Blodgett assembly of graphite oxide single layers [J]. Journal of the American Chemical Society, 2009, 131 (3): 1043-1049.

[58] Li Xiaolin, Zhang Guangyu, Bai Xuedong, et al. Highly conducting graphene sheets and Langmuir-

Blodgett films [J]. Nature Nanotechnology, **2008**, 3 (9): 538-542.

[59] Zheng Qingbin, Hing Ip Wai, Lin Xiuyi, et al. Transparent conductive films consisting of ultralarge graphene sheets produced by Langmuir-Blodgett assembly [J]. Acs Nano, **2011**, 5 (7): 6039-6051.

[60] Byon Hye Ryung, Lee Seung Woo, Chen Shuo, et al. Thin films of carbon nanotubes and chemically reduced graphenes for electrochemical micro-capacitors [J]. Carbon, **2011**, 49 (2): 457-467.

[61] Kumarasamy Jayakumar, Camarada María Belén, Venkatraman Dharuman, et al. One-step coelectrodeposition-assisted layer-by-layer assembly of gold nanoparticles and reduced graphene oxide and its self-healing three-dimensional nanohybrid for an ultrasensitive DNA sensor [J]. Nanoscale, **2018**, 10 (3): 1196-1206.

[62] Kim Jaemyung, J Cote Laura, Kim Franklin, et al. Graphene oxide sheets at interfaces [J]. Journal of the American Chemical Society, **2010**, 132 (23): 8180-8186.

[63] Chen Long, Huang Liangliang, Zhu Jiahua. Stitching graphene oxide sheets into a membrane at a liquid/liquid interface [J]. Chemical Communications, **2014**, 50 (100): 15944-15947.

[64] Shen Bin, Zhai Wentao, Zheng Wenge. Ultrathin Flexible Graphene Film: An Excellent Thermal Conducting Material with Efficient EMI Shielding [J]. Advanced Functional Materials, **2014**, 24 (28): 4542-4548.

[65] Becerril Héctor A., Mao Jie, Liu Zunfeng, et al. Evaluation of solution-processed reduced graphene oxide films as transparent conductors [J]. Acs Nano, **2008**, 2 (3): 463-470.

[66] M. Beese Allison, An Zhi, Sarkar Sourangsu, et al. Defect-Tolerant Nanocomposites through Bio-Inspired Stiffness Modulation [J]. Advanced Functional Materials, **2014**, 24 (19): 2883-2891.

[67] Wang Jie, Liang Minghui, Fang Yan, et al. Rod-coating: towards large-area fabrication of uniform reduced graphene oxide films for flexible touch screens [J]. Advanced Materials, **2012**, 24 (21): 2874-2878.

[68] Yulong Ying, Ying Wen, Guo Yi, et al. Cross-flow-assembled ultrathin and robust graphene oxide membranes for efficient molecule separation [J]. Nanotechnology, **2018**, 29 (15): 155602.

[69] Sun Hongyan, Chen Ding, Ye Chen, et al. Large-area self-assembled reduced graphene oxide/electrochemically exfoliated graphene hybrid films for transparent electrothermal heaters [J]. Applied Surface Science, **2018**, 435: 809-814.

[70] Yao Cen, Sun Yu, Zhao Kaisen, et al. Self-assembled layer-by-layer partially reduced graphene oxide-sulfur composites as lithium-sulfur battery cathodes [J]. RSC Advances, **2018**, 8 (7): 3443-3452.

[71] Xu Weiwei, Fang Chao, Zhou Fanglei, et al. Self-Assembly: A Facile Way of Forming Ultrathin, High-Performance Graphene Oxide Membranes for Water Purification [J]. Nano Letters, **2017**, 17 (5): 2928-2933.

[72] Kimal Chandula Wasalathilake, Dilini G. D. Galpaya, Godwin A. Ayoko, et al. Understanding the structure-property relationships in hydrothermally reduced graphene oxide hydrogels [J]. Carbon, **2018**, 137: 282-290.

[73] Umbreen Nadia, Sohni Saima, Ahmad Imtiaz, et al. Self-assembled three-dimensional reduced graphene oxide-based hydrogel for highly efficient and facile removal of pharmaceutical compounds from aqueous solution [J]. Journal of Colloid Interface Science, **2018**, 527: 356-367.

[74] Sasikala Suchithra Padmajan, Lim Joonwon, Kim In Ho, et al. Graphene oxide liquid crystals: a frontier 2D soft material for graphene-based functional materials [J]. Chemical Society Reviews, **2018**, 47 (16): 6013-6045.

[75] Marcano Daniela C, Kosynkin Dmitry V, Berlin Jacob M, et al. Improved synthesis of graphene oxide

[J]. Acs Nano, 2010, 4 (8): 4806-4814.

[76] Xu Yuxi, Sheng Kaixuan, Li Chun, et al. Self-assembled graphene hydrogel via a one-step hydrothermal process [J]. Acs Nano, 2010, 4 (7): 4324-4330.

[77] Zhang Li, Shi Gaoquan. Preparation of Highly Conductive Graphene Hydrogels for Fabricating Supercapacitors with High Rate Capability [J]. Journal of Physical Chemistry C, 2011, 115 (34): 17206-17212.

[78] Tang Zhihong, Shen Shuling, Zhuang Jing, et al. Noble-metal-promoted three-dimensional macroassembly of single-layered graphene oxide [J]. Angewandte Chemie, 2010, 49 (27): 4603-4607.

[79] Zhang Zheye, Xiao Fei, Guo Yunlong, et al. One-pot self-assembled three-dimensional TiO_2-graphene hydrogel with improved adsorption capacities and photocatalytic and electrochemical activities [J]. Acs Appl Mater Interfaces, 2013, 5 (6): 2227-2233.

[80] Wu Zhong-Shuai, Yang Shubin, Sun Yi, et al. 3D nitrogen-doped graphene aerogel-supported Fe_3O_4 nanoparticles as efficient electrocatalysts for the oxygen reduction reaction [J]. Journal of the American Chemical Society, 2012, 134 (22): 9082-9085.

[81] Zhao Yang, Liu Jia, Hu Yue, et al. Highly compression-tolerant supercapacitor based on polypyrrole-mediated graphene foam electrodes [J]. Advanced Materials, 2013, 25 (4): 591-595.

[82] Jiang Xu, Ma Yanwen, Li Juanjuan, et al. Self-Assembly of Reduced Graphene Oxide into Three-Dimensional Architecture by Divalent Ion Linkage [J]. Journal of Physical Chemistry C, 2010, 114 (51): 22462-22465.

[83] Wu Zhong-Shuai, Winter Andreas, Chen Long, et al. Three-dimensional nitrogen and boron co-doped graphene for high-performance all-solid-state supercapacitors [J]. Advanced Materials, 2012, 24 (37): 5130-5135.

[84] Huang Yanshan, Wu Dongqing, Han Sheng, et al. Assembly of tin oxide/graphene nanosheets into 3D hierarchical frameworks for high-performance lithium storage [J]. Chemsuschem, 2013, 6 (8): 1510-1515.

[85] Niu Zhiqiang, Liu Lili, Zhang Li, et al. A universal strategy to prepare functional porous graphene hybrid architectures [J]. Advanced Materials, 2014, 26 (22): 3681-3687.

[86] Shi Ji Lei, Du Wen Cheng, Yin Ya Xia, et al. Hydrothermal reduction of three-dimensional graphene oxide for binder-free flexible supercapacitors [J]. Journal of Materials Chemistry A, 2014, 2 (28): 10830-10834.

[87] Liu Xin, Cui Jiashan, Sun Jianbo, et al. 3D graphene aerogel-supported SnO_2 nanoparticles for efficient detection of NO_2 [J]. Rsc Advances, 2014, 4 (43): 22601-22605.

[88] Ma Jingjing, Wang Jiulin, He Yushi, et al. A Solvothermal Sorategy one step in situ synthesis of self assembled 3D graphene-based composites with enhanced lithium storage capacity [J]. Journal of Materials Chemistry A, 2014, 2 (24): 9200-9207.

[89] Sheng Kai Xuan, Yu-Xi X. U., Chun L. I., et al. High-performance self-assembled graphene hydrogels prepared by chemical reduction of graphene oxide [J]. New Carbon Materials, 2011, 26 (1): 9-15.

[90] Moon In, Junghyun Lee, Ruoff Rodney S, et al. Reduced graphene oxide by chemical graphitization [J]. Nature Communications, 2010, 1 (6): 73.

[91] Park Sungjin, An Jinho, Potts Jeffrey R., et al. Hydrazine-reduction of graphite and graphene oxide [J]. Carbon, 2011, 49 (9): 3019-3023.

[92] Chen Wufeng and Yan Lifeng. In situ self-assembly of mild chemical reduction graphene for three-dimensional architectures [J]. Nanoscale, 2011, 3 (8): 3132-3137.

[93] Fan Zeng, Tng Daniel Zhi Yong, Nguyen Son Truong, et al. Morphology effects on electrical and thermal

properties of binderless graphene aerogels [J]. Chemical Physics Letters，2013，561-562（3）：92-96.

[94] Gao Hongcai，Xiao Fei，Bun Ching Chi，et al. High-performance asymmetric supercapacitor based on graphene hydrogel and nanostructured MnO_2 [J]. Acs Applied Materials Interfaces，2012，4（5）：2801-2810.

[95] Zhang Lianbin，Chen Guoying，Hedhili M，et al. Three-dimensional assemblies of graphene prepared by a novel chemical reduction-induced self-assembly method [J]. Nanoscale，2012，4（22）：7038-7045.

[96] Adhikari Bimalendu，Biswas Abhijit，Banerjee Arindam. Graphene oxide-based hydrogels to make metal nanoparticle-containing reduced graphene oxide-based functional hybrid hydrogels [J]. Acs Applied Materials Interfaces，2012，4（10）：5472-5482.

[97] Worsley Marcus A，Pauzauskie Peter J，Olson Tammy Y，et al. Synthesis of graphene aerogel with high electrical conductivity [J]. Journal of the American Chemical Society，2010，132（40）：14067-14069.

[98] Wang Zhong-Li，Xu Dan，Xu Ji-Jing，et al. Graphene Oxide Gel-Derived，Free-Standing，Hierarchically Porous Carbon for High-Capacity and High-Rate Rechargeable $Li-O_2$ Batteries [J]. Advanced Functional Materials，2012，22（17）：3699-3705.

[99] Yang Jinli，Wang Jiajun，Wang Dongniu，et al. 3D porous $LiFePO_4$/graphene hybrid cathodes with enhanced performance for Li-ion batteries [J]. Journal of Power Sources，2012，208（2）：340-344.

[100] Xu Yuxi，Wu Qiong，Sun Yiqing，et al. Three-dimensional self-assembly of graphene oxide and DNA into multifunctional hydrogels [J]. Acs Nano，2010，4（12）：7358-7362.

[101] Wan Wubo，Li Lingli，Zhao Zongbin，et al. Ultrafast Fabrication of Covalently Cross-linked Multifunctional Graphene Oxide Monoliths [J]. Advanced Functional Materials，2015，24（31）：4915-4921.

[102] Wang Chao，Duan Yipin，S Zacharia Nicole，et al. A family of mechanically adaptive supramolecular graphene oxide/poly（ethylenimine）hydrogels from aqueous assembly [J]. Soft Matter，2017，13（6）：1161-1170.

[103] Wang Yanqing，Fugetsu Bunshi，Sakata Ichiro，et al. Morphology-controlled fabrication of a three-dimensional mesoporous poly（vinyl alcohol）monolith through the incorporation of graphene oxide [J]. Carbon，2016，98：334-342.

[104] Wang Kaili，Wang Wei，Wang Haibo，et al. 3D graphene foams/epoxy composites with double-sided binder polyaniline interlayers for maintaining excellent electrical conductivities and mechanical properties [J]. Composites Part A Applied Science Manufacturing，2018，110：246-257.

[105] Huang Xiaodan，Zhao Yufei，Ao Zhimin，et al. Micelle-template synthesis of nitrogen-doped mesoporous graphene as an efficient metal-free electrocatalyst for hydrogen production [J]. Sci Rep，2014，4：7557.

[106] Huang Xiaodan，Sun Bing，Su Dawei，et al. Soft-template synthesis of 3D porous graphene foams with tunable architectures for lithium-O_2 batteries and oil adsorption applications [J]. Journal of Materials Chemistry A，2014，2（21）：7973-7979.

[107] Zhang Fei Fei，Zhang Xin Bo，Dong Yun Hui，et al. Facile and effective synthesis of reduced graphene oxide encapsulated sulfur via oil/water system for high performance lithium sulfur cells [J]. Journal of Materials Chemistry，2012，22（23）：11452-11454.

[108] Li Yingru，Chen Ji，Huang Liang，et al. Highly compressible macroporous graphene monoliths via an improved hydrothermal process [J]. Advanced Materials，2014，26（28）：4789-4793.

[109] Barg Suelen，Macul Perez Felipe，Ni N，et al. Mesoscale assembly of chemically modified graphene into complex cellular networks [J]. Nature Communications，2014，5：4328.

[110] Menzel Robert，Barg Suelen，Miranda Miriam，et al. Joule Heating Characteristics of Emulsion-Tem-

[111] Zhang Rujing, Cao Yachang, Li Peixu, et al. Three-dimensional porous graphene sponges assembled with the combination of surfactant and freeze-drying [J]. Nano Research, **2014**, 7 (10): 1477-1487.

[112] Su DangáSheng. Macroporous "bubble" graphene film via template-directed ordered-assembly for high rate supercapacitors [J]. Chemical communications, **2012**, 48 (57): 7149-7151.

[113] Niu Zhiqiang, Chen Jun, Hng Huey Hoon, et al. A leavening strategy to prepare reduced graphene oxide foams [J]. Advanced Materials, **2012**, 24 (30): 4144-4150.

[114] El-Kady Maher, Strong Veronica, Dubin Sergey, et al. Laser scribing of high-performance and flexible graphene-based electrochemical capacitors [J]. Science, **2012**, 335 (6074): 1326-1330.

[115] Strong Veronica, Dubin Sergey, El-Kady Maher, et al. Patterning and electronic tuning of laser scribed graphene for flexible all-carbon devices [J]. Acs Nano, **2012**, 6 (2): 1395-1403.

[116] Hao Fatt Teoh, Ye Tao, Tok Eng Soon, et al. Direct laser-enabled graphene oxide-Reduced graphene oxide layered structures with micropatterning [J]. Journal of Applied Physics, **2012**, 112 (6): 206-327.

[117] Sibel Korkut, Roy-Mayhew Joseph D, Dabbs Daniel M, et al. High surface area tapes produced with functionalized graphene [J]. Acs Nano, **2011**, 5 (6): 5214-5222.

[118] Wang Chunhui, Chen Xiong, Wang Bin, et al. Freeze-Casting Produces a Graphene Oxide Aerogel with a Radial and Centrosymmetric Structure [J]. ACS nano, **2018**, 12 (6): 5816-5825.

[119] Wu Zhong-Shuai, Shi Xiaoyu, Xiao Han, et al. Graphene-based Porous Materials for Advanced Energy Storage in Supercapacitors, in Innovations in Engineered Porous Materials for Energy Generation and Storage Applications [M]. CRC Press, 2018: 59-85.

[120] Teradal Nagappa L., Marx Sharon, Morag Ahiud, et al. Porous graphene oxide chemi-capacitor vapor sensor array [J]. Journal of Materials Chemistry C, **2017**, 5 (5): 1128-1135.

第5章
CHAPTER 5

氧化石墨烯基复合材料

GO 具有丰富的官能团，便于改性修饰，相比于石墨烯更加容易分散到复合材料当中，且能够保持其片状结构。经过还原过程还原氧化石墨烯后，可将石墨烯的众多优点赋予复合材料，实现复合材料性能的提升。因此 GO 基复合材料研究受到了广泛关注。

5.1 氧化石墨烯/高分子复合材料体系

纳米粒子作为填料制备的高分子复合材料具有优异的性能，广泛应用于汽车、飞机、建筑、电子器件等领域。其中性能的提升与纳米粒子在复合材料中的分散状态和纳米粒子与高分子基体之间的相互作用有很大的关系[1~4]。多数纳米粒子与高分子不相容，在复合材料中无法形成均相体系，从而制约了纳米粒子对高分子复合材料的增强作用[5,6]。GO 表面具有丰富的官能团，与很多高分子材料之间有较高相容性，可以用作多种高分子复合材料增强填料，得到的复合材料可以提升力学、电学、热学等多方面性能[7~9]。

5.1.1 制备方法

常见的制备方法有原位聚合法、溶液共混法以及熔融共混法。

原位聚合法是利用 GO 表面的羟基、环氧基、羧基等官能团，与高分子进行化学键合同时进行聚合。根据反应过程，可分为 3 类：①从 GO 表面开始，原位聚合制备 GO-高分子材料（Graft-from GO）；②将制得的高分子链与 GO 链接（Graft-onto GO）；③GO 与单体混合后，在原位聚合过程中分散到高分子体系里面。

Graft-from GO 的反应，通常应用于可控活性聚合反应，单体从 GO 表面开始逐步聚合得到高分子材料。这种方法反应效率高，既可直接制备 GO 增强的复合材料，产物还可以作为改性填料用于其他类似性质高分子材料的增强。这种方法可以制得较为规整的高分子刷结构，对高分子本身的结构性质影响较小，可用作高分子受限结晶研究的物理模型。但是，由于 GO 表面的官能团羟基、环氧基和羧基不能引发活性聚合，所以经常需要先对 GO 进行化学修饰，常见的有接枝 Ziegler-Natta 催化剂，以制备聚烯烃（如 GO-g-PP）复合材料[10]；接枝 ATRP 聚合引发前体，以制备聚丙烯酸类、聚苯乙烯类复合材料[11~13]；制备 RAFT 聚合引发前体，以引发乙烯基类单体聚合反应（如 GO-g-PVK）[14]。

在 Graft-onto GO 方法中，先合成高分子以避免 GO 对高分子聚合的影响，但是接枝效率偏低。一般的思路是利用高分子的端基与 GO 表面官能团之间反应，常见的有端羟基与 GO 羧基酯化反应[15]、端氨基与 GO 羧基反应[16~19]、硅烷偶联剂与 GO 羟基反应[20] 等。为提高接枝效率，可先活化 GO 表面的羟基、羧基基团，然后将其与高分子的活性端基反应，接枝成 GO 改性高分子材料，常见的有前述的 click 点击反应[21,22]。

将 GO 或改性 GO 与单体混合，直接原位聚合的方法相对简单，可以大规模制备 GO 改性复合材料。但是这种反应接枝率不可控。为了提高反应效率，可对 GO 进行部分改性，以提高其反应活性，通常应用于环氧树脂、聚氨酯等热固性树脂体系的合成[23~30]。另外也有研究直接利用 GO 表面产生自由基，与丙烯酸类、苯乙烯类、丁二烯类等单体自由基聚合，制备 GO 高分子复合材料[31~35]，但是这种方法分子结构较难控制，产物多用作填料或构建交联体系。

由于 GO 自身为刚性片层状结构，与高分子的链接结构几何构型不匹配，因此在复合材料中经常会产生相分离。为抑制 GO 在复合材料中的分相行为，可通过聚合将 GO 与高分子链化学键合或对其进行相应的改性[36]。通过该方法，一方面可提高 GO 与复合材料的相容性，有利于 GO 各项性能的发挥，减少 GO 用量；另一方面可提高 GO 在不同溶剂中的分散性，使其作为填料应用于不同体系中，满足多数溶液加工的要求。

溶液共混法是将 GO 或改性 GO 与高分子在溶剂中混合，然后除掉溶剂得到复合材料。该方法加工简单，通过搅拌或研磨即可将 GO 分散到高分子体系中。然而，为保证分散效果，该方法通常要使用大量有机溶剂，环保压力较大。因此，该方法一般多应用于环氧、聚氨酯、丙烯酸树脂等涂料体系的生产加工，对于塑料、橡胶等复合材料体系，多应用于实验室样品制备，生产中应用较少。

熔融共混法是在高温下将 GO 或改性 GO 与高分子熔体直接混合制备复合材料。通常使用螺杆、混炼机等设备将材料混合后，再挤出、注塑等方法成型。该方法操作简单，广泛应用于塑料、橡胶等产品的生产加工。其缺点则是混合时体系黏度较大、混合时间短，从而导致石墨烯分散性较差且较难控制。

5.1.2 应用

GO 加入高分子复合材料通常有如下应用目的：a.提升材料的力学性能；b.提升材料导电性；c.提升材料阻隔性能；d.提升材料的热稳定性和阻燃能力；e.调节材料能级，应用于光电材料；f.制备高分子刷，研究高分子结晶行为等。

5.1.2.1 提升力学性能

利用 GO 提升复合材料的力学性能是 GO 一个主要应用，关键是提高 GO 在复合材料中的分散性和调控 GO 与高分子基体间的相互作用[37]。一般而言，加入 GO 可以显著增强复合材料的强度与韧性，且 GO 与高分子基体相容性越好，增强效果越明显；反之则效果降低，甚至会降低材料的韧性。由于 RGO 官能团较少，在增强复合材料强度的同时通常降低韧性。

不同的添加方式会导致不同的效果。原位聚合的方法既可以提高 GO 在高分子基体中的分散性，又能保证 GO 与高分子基体之间较好的化学键合；溶液共混法制备的复合材料中，GO 分散性较好，但界面较难调控；熔融共混法中 GO 较难分散并不容易控制界面，得到的复合材料性能不易控制。表 5.1 总结了利用溶液共混方式在不同高分子基体中添加各种 GO 时的力学性能提升情况，以供参考。

值得注意的是，通过溶液交换法可以提高 GO 聚合物类复合材料的力学稳定性，有效地提升复合材料的韧性、抗拉伸性以及杨氏模量。研究人员通过反复实验得出结果，在壳聚糖基体中的 GO 可以实现良好分散，进而有效提升这一复合材料的力学性能[38]。

表 5.1 不同类型 GO 在不同基体中的力学增强效果

基体	填料类型	添加量/[%(质量)[①], %(体积)[②]]	加工方式	性能提升	参考文献
环氧	改性 GO	2%[①]	溶液共混	储能模量提高 113%	[23]
		4%[①]	溶液共混	硬度提升 38%	[23]
聚氨酯	异氰酸酯改性 GO	3%[①]	溶液共混	抗拉刚度提升 10 倍	[26]
	异氰酸酯改性 RGO	2%[①]	原位聚合	拉伸强度提升 239%，储能模量提升 202%	[28]
聚苯并咪唑	GO	0.3%[①]	溶液共混	杨氏模量提高 17%，拉伸强度提高 33%，韧性提高 88%	[39]

续表

基体	填料类型	添加量/[%(质量)①,%(体积)②]	加工方式	性能提升	参考文献
壳聚糖	GO	1%①	溶液共混	拉伸强度从40.1MPa提高到89.2MPa,杨氏模量由1.32GPa提升至2.17GPa	[38]
PVA	GO	0.7%①	溶液共混	断裂强度由49.9MPa增加到87.6MPa	[40]
PMMA	RGO	1%①	溶液共混	弹性模量提高28%,断裂强度下降8.8%	[31]
PMMA	GO	1%①	溶液共混	弹性模量提高28%,断裂强度提高22.0%	[31]
PMMA	GO-g-PMMA	1%①	溶液共混	断裂伸长率提高1倍,杨氏模量提高16%	[11]
PMMA	GO	1%①	溶液共混	杨氏模量降低22%	[11]
PS	GO-SEBS	2%①	溶液共混	拉伸强度提高78%,杨氏模量提高73%	[21]
SBR	GO-XNBR乳液	3%①	溶液共混	拉伸强度提高332%	[41]
SBR	GO交联SBR	3%①	溶液共混	拉伸强度提高4倍	[34]
SBR	油胺改性GO	7%①	溶液共混	储能模量提高67%	[42]
SBR	十八胺改性GO	7%①	溶液共混	储能模量提高39%	[42]
SBR	聚乙烯吡咯烷酮改性GO	5%①	溶液共混	提高拉伸强度517%,撕裂强度387%	[43]
SBR	GO	5%①	溶液共混	拉伸强度提高332%,撕裂强度提高277%	[43]
NR	BTESPT改性GO	0.3%①	溶液共混	拉伸强度提高100%,拉伸模量提高66%	[44]
VPR	氨基改性GO	3.6%②	溶液共混	玻璃态模量提高21倍,橡胶态模量提高7.5倍,拉伸强度提高3.5倍	[45]

① 质量分数。
② 体积分数。
注: SBR, 丁苯橡胶, Polymerized Styrene Butadiene Rubber; NR, 天然橡胶, Natural Rubber; VPR, 乙烯基聚酯树脂, Vinyl Polyester Resin; SEBS, 是以聚苯乙烯为末端段、以聚丁二烯加氢得到的乙烯-丁烯共聚物为中间弹性嵌段的线性三嵌共聚物。

Wang等[39]人用溶液交换法制备出了GO/聚苯并咪唑复合材料。研究发现,仅加入0.3%的GO就使复合材料的杨氏模量提高了17%,拉伸强度提高了33%,韧性提高了88%。与单纯的聚苯并咪唑相比,复合材料的杨氏模量、拉伸强度以及韧性均得到显著改善,热稳定性也有所提高,而且成本相对低廉,有效解决了聚苯并咪唑价格昂贵的问题。此外,将壳聚糖和GO在水溶液中自组装,也制备出了GO/壳聚糖纳米复合材料,其中GO在壳聚糖基体中分散良好,并且彼此有相互作用。拉伸实验结果表明，加入 1% GO 后，拉伸强度从 40.1MPa 提高到 89.2MPa，杨氏模量由 1.32GPa 提升至 2.17GPa。Liang 等[40]人制备了 PVA/GO 复合材料，GO 含量为 0.7% 时，复合材料的断裂强度由纯 PVA 的 49.9MPa 增加到 87.6MPa。

将 GO 改性之后与环氧树脂原位聚合得到 GO-环氧树脂复合材料，2% 的 GO 添加量就可使复合材料储能模量提高 113%；4% 的 GO 添加量可使复合材料硬度提升 38%，且 GO 在树脂热固化的过程中部分还原，提高了材料的导电性，使其面电阻降低 6.5 个数量级[23]。

在聚氨酯体系中，热还原的 GO（记作 TGO）与异氰酸酯改性的 GO（记作 iGO）添加表现出的性能不同。溶液共混的条件下，TGO 改性的聚氨酯导电性能最好，0.5%（质量）的添加量即可使表面电阻降到 $10^5\Omega$ 的水平；而 iGO 则在屏蔽和增强方面表现突出，3%（质量）的添加量可使复合材料的抗拉刚度提升 10 倍，氮气透过率降低 90%。使用熔融共混的加工方式，增强作用明显减弱，这是因为熔融共混条件下 TGO 和 iGO 分散程度较差。使用 TGO 与聚氨酯单体原位聚合制备的复合材料性能也差于溶液共混的 iGO-聚氨酯复合材料[26]。相反地，将化学还原的 GO 用异氰酸酯改性，然后再与聚氨酯单体原位聚合，得到的复合材料力学性能、导电性能和热稳定性均大幅提升，2%（质量）的 RGO 用量即可使复合材料拉伸强度提升 239%，储能模量提升 202%[28]。这说明，只有形成有效的化学接枝，原位聚合方式制备的复合材料性能才会明显提升，进一步说明了提高 GO 与高分子复合材料相容性的重要性。

将 RGO 与 GO 作为填料，分别加入到甲基丙烯酸甲酯（Methyl Methacrylate，简称 MMA）单体中可制备 PMMA/石墨烯复合材料。相比于纯 PMMA，添加 1%（质量）的 RGO 或 GO，均可使复合材料的弹性模量提高 28%。然而，RGO 使复合材料的断裂强度下降 8.8%，而 GO 则使复合材料的断裂强度提高 22.0%，说明 GO 与 PMMA 的界面相容性更好[31]。为进一步提高 GO 与 PMMA 的接枝效率，还可采用 ATRP 聚合方法合成 GO-g-PMMA 复合材料，实现在 PMMA 中均匀分散，大幅提高 PMMA 的韧性。由于制备工艺相对复杂，GO-g-PMMA 常用作增强填料。1%（质量）的 GO-g-PMMA 添加量，可使 PMMA 的断裂伸长率提高 1 倍，杨氏模量提高 16%；同等条件下，添加 1%（质量）的 GO 则对 PMMA 断裂伸长率几乎没影响，杨氏模量降低 22%[11]。

通过点击方法合成的 GO-SEBS 复合材料，可结合 GO 的增强性能与 SEBS 的增韧性能，作为填料可同时提高复合材料的强度与韧性。例如，将 2% 的 GO-SEBS 加入到 PS 中，复合材料的拉伸强度可相比于未添加的 PS 提高 78%，杨氏模量提高 73%，而且 10% 热分解温度由 325℃ 提升至 351℃，热稳定性明显提高[21]。

在橡胶类体系中，需要同时兼顾材料的强度与韧性，因此对 GO 的分散性和 GO 与橡胶基体间的相互作用要求更高。主要通过将 GO 与橡胶分子交联，或对 GO 改性，增强其对橡胶分子的亲和性来实现[46,47]。Liu 等[41]人以极性 xNBR 为载体，将 GO 转移到 SBR 基体中。GO 悬浮液与 xNBR 胶乳混合，然后将其加入到 SBR 胶乳中，再进行胶乳共凝聚。用 XRD 和 SEM 对填料在 SBR 基体中的分散进行了表征，并研究了纳米复合材料的力学性能。研究发现，xNBR 可以通过氢键与 GO 相互作用，并与 SBR 形成化学交联。因此 xNBR 可以防止 SBR 基体中 GO 片层聚集，改善 GO 和 SBR 的相互作用。

与纯 SBR 相比，添加 1 份 xNBR-GO 可以使 SBR 的拉伸强度提高 332%，而添加 1 份 GO 仅能使 SBR 拉伸强度提高 185%。当添加 3 份 xNBR-GO 时，其对 SBR 拉伸强度的增强作用达到最大，约 900%；而单纯添加 GO 要 5 份才达到约 500%。这说明了经过 xNBR 改性，提高了 GO 与 SBR 之间的相容性，从而提高了其分散性与增强能力。

GO 在高温下能产生自由基，可直接促使橡胶分子的交联。因此可以使用 GO 代替硫或过氧化二异丙苯使橡胶交联，同时 GO 还可作为填料增强橡胶的力学性能。相比于利用硫或过氧化二异丙苯的交联方式，使用 GO 交联的 SBR 拉伸强度提高 4 倍。该方法的缺点是交联固化所需的时间过长，通常需要 1 小时以上的时间才能交联完[34]。

用双-[3-(三乙氧基硅)丙基]-四硫化物（BTESPT）的硅氧键与 GO 的羟基反应，合成 BTESPT 改性的 GO（SGO）。将其加入橡胶中后，其中的硫与橡胶分子交联，实现 GO 与橡胶交联，可以同时提高橡胶材料的力学性能和气体阻隔性能。0.3%（质量）的 SGO 即可使 NR 的拉伸强度提高 100%，拉伸模量提高 66%，同时透气性降低 48%[44]。

用油胺与十八胺对 GO 进行改性，然后与 SBR 溶液混合均匀，共凝聚后制得改性 GO-SBR 复合材料。无论在玻璃态和橡胶态，改性的 GO-SBR 与纯 GO-SBR 相比储能模量均大幅提高；25℃时，7%（质量）油胺改性 GO 和 7%（质量）十八胺改性 GO 分别使橡胶储能模量提高了 67% 和 39%。这其中主要的原因是氨基改性的 GO 相比于纯 GO 在 SBR 中分散性更好，且与橡胶界面作用更强。两种胺之间的性能区别主要是：油胺含有双键，在硫化过程中可以与橡胶交联，从而进一步提高橡胶性能[42]。同样的现象在丁二烯-苯乙烯-乙烯基吡啶橡胶中也被观察到。在 VPR 中添加 3.6%（体积）的氨基改性 GO，可以使复合材料的玻璃态模量提高 21 倍，橡胶态模量提高 7.5 倍，拉伸强度提高 3.5 倍[45]。

聚乙烯吡咯烷酮可以与 GO 形成氢键作用，制备改性 GO（记作 PGO）用以增强橡胶体系。GO 与 PGO 均能提高橡胶的性能，但 PGO 效果更加显著。添加 5 份 PGO 到 100 份 SBR 中，可以将 SBR 的拉伸强度提高 517%，撕裂强度提高 387%；而同等量未改性的 GO 只能使 SBR 拉伸强度提高 332%，撕裂强度提高 277%。此外 PGO 还使复合材料的最高分解温度提升 23.6℃，增强了其热稳定性；减少甲苯吸收量 41%，提高了其耐溶剂能力；提升其热传导能力 30%。而 GO 只能使复合材料热传导能力提升 17%。可以看出，PGO 全面提升了 SBR 的各种性能，主要原因和 PGO 在橡胶基体中分散均匀有关,而且聚乙烯吡咯烷酮较大改善了橡胶分子与 GO 之间的相容性,使 GO 的增强性能得到充分发挥。继续增加填料的用量,GO/SBR 复合材料的性能开始下降,而 PGO/SBR 复合材料的性能还可继续提升。原因是增加填料用量时,GO 由于分散性不佳,开始发生严重团聚,从而导致材料性能下降[43]。

5.1.2.2 提升电学性能

单纯的导电聚合物在充放电循环的过程中通常稳定性较差,使得其在电容器电极等方面的应用受到了限制,开发具有优异导电性能的复合材料势在必行。石墨烯和导电聚合物共轭结构的相互作用可以增强基体导电性,同时又可以实现结构的增强。因此,导电聚合物与 GO 的复合成为一个研究热点[48]。虽然 GO 本身并不导电,但是在高分子加工过程中可以部分还原 GO,而导电填料与基体间的强界面作用以及导电填料在基体中良好的分散性能更有利于聚合物基体导电性能的提高[49]。表 5.2 列出了 GO 在一些类型的高分子基体中电学性能提升效果。

表 5.2　GO 在不同基体中的电学增强效果

基体	添加量/[%(质量)①,%(体积)②]	加工方式	性能提升	参考文献
聚苯乙烯(PS)	2%①	原位聚合	导电率 2.9×10^{-2} S/cm	[50]
PP	1.52%①	原位聚合	导电率 10^{-6} S/m	[10]
聚对苯二甲酸乙二醇酯	3%②	熔融共混	导电率 2.11S/cm	[51]
聚苯乙烯	0.38%②	熔融共混	导电率 5.77S/cm	[52]
PVC	10%①	熔融共混	电阻率 $10^3\Omega\cdot$cm	[53]
SBS	0.12%②	熔融共混	导电渗流阈值低至 0.12%(体积)	[54]
环氧	4%①	溶液共混	表面电阻降低 6.5 个数量级	[23]

① 质量分数。
② 体积分数。
注:PVC,聚氯乙烯,polyvinyl chloride。

Hu 等[50]人利用原位聚合法制备了 GO/聚苯乙烯导电复合材料,结果发现当石墨烯含量为 2%(质量)时,复合材料的导电率达到最高 2.9×10^{-2} S/cm,认为 GO 在基体中分散性较好且能形成有效的导电网络。

用格氏试剂将 GO 表面的羟基、环氧基和羧基格氏化,然后与 $TiCl_4$ 反应可制备 Ziegler-Natta 催化剂。利用改性过的催化剂,原位催化丙烯在 GO 表面聚合可生成聚丙烯-g-GO(PP-g-GO)复合材料[10]。该复合材料在 PP 树脂中可均匀分散,减少 GO 在 PP 中的团聚。PP-g-GO 在高温(190℃)加工过程中,GO 被初步还原,从而提高了复合材料的导电性。通过这种原位聚合的方式,1.52%(质量)的

GO 添加量即可使复合材料达到导静电的水平（10^{-6} S/m）。

Zhang 等[51]人以 GO 为填料，利用熔融共混法制备了石墨烯/聚对苯二甲酸乙二醇酯导电复合材料，结果发现石墨烯在基体中分散均匀，加工过程中 GO 被还原，有效地提高了基体的导电性能；当石墨烯的体积分数达到 3.0% 时，复合材料的导电率可达到 2.11S/cm。程博等人[53]利用熔融共混法制备了 GO/PVC 复合材料，结果发现石墨烯作为导电填料能够明显提高 PVC 的抗静电性；当其含量到达渗流阈值时［10%（质量）］，复合材料的体积电阻率到达最低值 $10^3\Omega\cdot cm$，与纯 PVC 相比体积电阻率降低了 12 个数量级。

Li 等[54]人制备了 GO/SBS 复合材料，结果发现 GO 在基体中具有良好的分散性，并且 GO 和基体之间的界面作用很强，从而在还原后提高了复合材料的导电性，导电渗流阈值低至 0.12%（体积）。陈翔峰等人[55]制备了 GO/丙烯腈苯乙烯导电复合材料，发现 GO 的径厚比对复合材料的体积电阻率有很大影响，径厚比大能够使其在基体中更易形成导电网络，从而降低复合材料的电阻率。

此外，不同的加工方式也会导致材料性能差异。通过研究对比三种方式制备的 RGO-PMMA 复合材料：a.原位聚合法直接制备 RGO-PMMA 复合材料；b.原位聚合 RGO-PMMA 预聚体与 PMMA beads 共混制备 RGO-PMMA 复合材料；c.原位聚合 RGO-PMMA 预聚体薄膜流延法制备 RGO-PMMA 复合材料。在 2%（质量）的 RGO 添加量下，薄膜流延方式制备的复合材料导电性提高了 8 个数量级，而 PMMA beads 共混合原位聚合得到的复合材料分别使导电性提高了 7 和 6.5 个数量级[32]。

此外，其他方面的应用也和聚合物导电性的提升紧密相关。例如，应用原位聚合法可以将 GO 与导电聚合物材料进行复合。这一方法可以在保证制备得到的超级电容器电极高充放电性能和高稳定性的同时提升电容器的安全性。聚合物和 GO 复合材料已经被广泛应用于电容器电极材料中，制备的电容器电极材料的比电容可达 421.4F/g 甚至更高[56~58]。

因此，还原后的 GO 作为填料对提升聚合物的导电性能具有明显的效果，极大地促进了各种高分子材料在多种电子元件生产中的应用。

5.1.2.3 提升阻隔性能

使用高阻隔性能高分子薄膜，可防止由于氧气等气体的渗透而引起的微生物繁殖和封装内容的氧化；防止香味、溶剂等的流出，提高内容物的储存性。进一步提高薄膜阻隔性能十分必要。高阻隔性包装材料如乙烯-乙烯醇共聚物、聚偏氯乙烯、聚胺、聚对苯二甲酸乙二醇酯等与 GO 复合，可使复合材料的阻隔性能得到进一步提升。

Wu 等[44]人报道了表面活性官能化的 GO（SGO）与双（三乙氧基硅丙基）四硫化物作为 NR 的多功能纳米填料的研究结果。研究者通过简单的方法成功地将双（三乙氧基硅丙基）四硫化物分子接枝到 GO 的表面上，得到的 SGO 可以通过溶液混合在 NR 中实现充分分散。研究发现，在低填充量下，SGO 显著地改善了 NR 的气体阻隔性能。研究人员将在 25℃时测量的 SGO/NR 纳米复合材料（记作 P）的透气性，与未填充 NR 的材料（记作 P_0）进行比较，对比 P/P_0 的值随 SGO 添加量变化。当 SGO 含量为 0.3%（质量）时，P/P_0 急剧下降至 52%，此后缓慢下降。根据实验，0.3%（质量）的 SGO 可与 16.7%的黏土添加效果相媲美，大幅度改善 NR 的气体阻隔性能。

实验证明，只有分散均匀的、具有高纵横比的片状结构材料才能提高复合材料的屏蔽能力，这是因为均匀分散的片状材料在复合材料中形成"迷宫效应"，增加了气体的扩散路径且减少了扩算截面积。表面官能化的 GO，自身屏蔽性高，且具有高纵横比、易分散等优点，在改善各种复合材料的气体阻隔性能方面可望有广泛应用。

5.1.2.4 提升热稳定性能

GO 在聚合物基体中可以限制聚合物链的流动性。在燃烧过程中，各向异性 GO 形成碳层网络，阻碍降解产物的逸出。还原后石墨烯还具有较高热导率，有助于燃烧区域聚集的热量扩散，因此 GO/聚合物复合材料可用作阻燃材料。GO 还可提高 PS、聚乙烯醇、聚甲基丙烯酸甲酯、聚氨酯等聚合物的耐热性[59,60]。这是因为 GO 的含氧基团与聚合物的氢键配位后，使复合材料的自由离子量缩减，进而在一定程度上降低了复合材料的振动频率。研究人员通过共混法，以 GO 和混合材料树脂用作原材料，进行 GO 聚合物复合材料的制备。实验结果发现，所制备的复合树脂材料与单纯的树脂相比，耐热性能有了显著的提升，这无疑为耐热材料的良好应用打下了坚实稳定的基础，也推动了耐热材料的发展[61]。

Xiong 等[61]人证明了溴化丁基橡胶（Butyl Bromide，简称 BIIR）的热稳定性可以使用 BIIR 与离子液体（Ionic Liquid，简称 ILs）改性后的 GO（GO-ILs）通过溶液共混来改善。研究人员对 GO-ILS/BIIR 纳米复合材料的热稳定性进行了系统分析和研究后发现，GO-ILS/BIIR 纳米复合材料的玻璃化转变温度（Glass Transition Temperature，简称 T_g）随着加入 GO-ILS 而升高，这表明 GO-ILS 与橡胶之间具有较强的界面黏合性。根据 TGA 结果也表明，在 BIIR 基体中掺入 GO-ILS 可以有效地提高橡胶纳米复合材料的热稳定性。利用基辛格方法对实验数据的分析，根据公式 $\ln\left(\dfrac{\beta}{T_{max}^2}\right) = \ln\left(\dfrac{AR}{g(\alpha)E_a}\right) - \dfrac{E_a}{RT_{max}}$ 对温度进行的拟合，得到的热分解活化能。可以看到，未填充 BIIR 的热分解活化能为 213kJ/mol，而在 GO-IL/BIIR 复合材料中对于 1%、2%、3%、4%含量的 GO-ILS，热分解活化能分别增加到 237kJ/mol、246kJ/mol、248kJ/mol、270kJ/mol。这表明复合材料的热稳定性得到了提高。

5.1.2.5 提升结晶性能

聚合物的结晶过程会直接影响其加工性能，GO 加入到聚合物中可以在复合体系中起到成核剂的作用，有效地改善聚合物的结晶过程。

研究人员对聚乳酸/GO 纳米复合材料进行了非等温和等温过程中冷结晶行为的研究[62]。通过不同升温速率的差热分析发现，随着 GO 负载量的增加，聚乳酸的结晶峰温向低温范围转移，说明聚乳酸的非等温冷结晶行为有明显改善，而且 GO 可显著地提高聚乳酸的结晶速率，并使得其结晶机理和晶体结构保持不变。

材料的结晶无疑与材料的性能和应用息息相关[63]。将 GO 与结晶材料复合，进而进行材料结晶过程的定向调整，可以实现材料性能的有效提升[64]。例如通过差热法研究发现，GO 的负载量在不断提升的同时，GO 聚合物的结晶现象也得到了有效的缓解。随着温度的不断降低，与原材料相比，GO 聚合物复合材料的结晶速度变得缓慢。与此同时，材料的基本结构并没有随着温度的降低而发生明显的改变。由此可见，一些 GO 聚合物复合材料可以被应用于各种低温环境当中，实现耐低温材料更加广泛的应用[65]。

5.1.2.6 提升导热性能

在工业上目前广泛使用的导热高分子材料有导热复合塑料、导热胶黏剂、导热涂层、导热覆铜板及各类导热橡胶及弹性体，如热界面弹性体等。目前复合型绝缘导热高分子主要是采用绝缘导热无机粒子如氮化硼、氮化硅和氧化铝等和聚合物基体复合而成；采用导体粒子和聚合物复合制备的导热聚合物，如碳材料、金属填充的导热高分子材料，适用于低绝缘或非绝缘导热场合。将 GO 同聚合物复合，其复合材料的导热性能大幅提升引起广泛关注。导热高分子主要应用于功率电子元器件、电机等设备的封装和电气绝缘及散热，和普通聚合物相比具有 4~10 倍的导热系数。

研究表明，GO 可以用于提高环氧树脂[66,67]、聚乙烯、聚酰胺等聚合物的导热性能。通常而言，碳基填料可以提高聚合物的导热系数，但无法像提高导电性那么明显，甚至低于有效介质理论[68]。其原因可能是因为热能传递主要是以晶格振动的形式，填料与聚合物之间以及填料与填料之间较弱的振动模式也会增加热阻。

液态硅橡胶（Liquid Silicone Rubber，简称 LSR）广泛应用于电子器件的密封。在一般情况下，LSR 的导热性较差使得涂层或盆栽器件散热过量，从而导致器件损坏或寿命降低。为了缓解这一现状，Mu 等[69] 人研究了宽体积范围内填充 ZnO 的硅橡胶的热导率，并研究了形成的导电粒子链对导热系数的影响；也研究了 Al_2O_3 用量对硅橡胶导热性能和力学性能的影响。研究发现，通过填充 ZnO、Al_2O_3 等填料，硅橡胶复合材料的导热系数有明显提升，但填料的用量却相当可观。此外，随着填料负载量的增加，硅橡胶复合材料的力学性能也越来越差，复合材料的黏度也越来越大，不利于进一步加工使用。为此，Zhao 等[70] 人研究了三乙氧基硅烷（Triethoxysilane，简称TEVS）偶联剂对GO的表面官能化作用，以提高LSR的性能。硅烷官能化的GO（TEVS-GO）可以大幅提高LSR复合材料的导热性能，并且对复合材料的黏度没有影响。在典型的实施过程中，TEVS-GO通过在甲苯中用超声波分散2小时，以获得均匀悬浮液。然后将悬浮液与聚甲基乙烯基硅氧烷（Polymethylvinylsiloxane，简称PMVS）（3∶1质量比）混合，在PMVS完全溶解后，将混合物在70℃搅拌至恒重，得到TEVS-GO/PMVS混合液。然后，将聚甲基氢硅氧烷（Polymethylhydrosiloxane，简称PMHS）、铂催化剂和抑制剂与TEVS-GO/PMVS混合物充分混合（PMVS∶PMHs＝1∶2的质量比）。在脱气去除气泡后，将混合物倒入聚四氟乙烯模具中，在80℃下固化24小时，通过上述步骤制备含有TEVS-GO质量比为0.05％、0.1％、0.15％、0.2％、0.25％和0.3％的TEVS-GO/LSR复合材料。

在25℃温度下不同填料复合材料的导热系数，随着填料的加入，聚合物复合材料的导热系数增加[70]。在添加量为0.3％的情况下，GO/LSR复合材料的导热系数达到0.3W/m·K，与LSR（0.19W/m·K）的初始导热系数相比，具有中等增强作用。然而，在相同情况下，TEVS-GO/LSR复合材料的导热系数为0.38W/m·K。TEVS-GO填料与未经修饰的GO相比只需要更小的填充量，具有更高的增强效率。当TEVS-GO填料含量从0.05％提高到0.3％时，TEVS-GO/LSR导热性能的提升从7.9％变成42.1％，这和TEVS-GO构成的导热网络有关。

5.1.2.7 提升不同聚合物间相容性

不同高聚物间共混可明显提升其各种物理性能，具有广阔的使用范围。通过改变聚合物的类型和组分的配比来调控聚合物共混物的性能，可以综合利用各组分的性能，是一种非常有效和经济的方法，从而满足各种应用的特定要求[71,72]。然而，简单的聚合物共混往往并不能满足性能要求，因为两种不相容的高聚物共混特别是混合焓比较大的共混胶，会发生明显的相分离[73]。

研究表明，GO表面具有疏水性基面和亲水性边缘[72,74]。这种两亲性使其与极性或非极性聚合物都能有效地相互发生作用，从而可以作为聚合物共混的融合剂[75~77]。例如，Cao等[72]人采用GO来增容聚乙酰胺/聚苯醚（PA/PPO，90/10）聚合物共混物，发现分散相（PPO相）液滴直径可减小1个数量级，表明PA/PPO共混物的相容性得到了提高。相比未添加GO的PA/PPO共混物，当加入重量比为1％的GO时，GO/PA/PPO的拉伸强度增加了87％。Yan等[77]人研究发现，GO添加到天然橡胶/高密度聚乙烯（High Density Polyethylene，简称HDPE）（NR/HDPE 60/40）共混物中，可以降低NR相的尺寸，并且改善两相之间的相容性和界面黏结。当GO用量为1.5份时，NR/HDPE共混物的拉伸强度和300％伸长率时拉伸强度（M300）分别增加了27％和24％。Chen等[78]人研究发现，添加重量比为0.5％的GO可以使三元乙丙橡胶/石油树脂（EPDM/PR，40/

60)共混物拉伸强度和断裂伸长率分别提高 50% 和 30%。

DSC 测试可进一步证实 GO 对羧基丁腈橡胶（Nipol Carboxy Lated Nitrile，简称 XNBR）/SBR 共混胶的增容效果[79]。研究发现，未填充填料的 XNBR/SBR 共混胶有两个 T_g，在 $-49.07℃$ 和 $-2.41℃$ 处，分别对应于 SBR 和 XNBR 的转变，说明 XNBR/SBR 共混胶的相容性不好。随着 GO 的加入，对应于 SBR 的 T_g 向高温方向移动，同时对应于 XNBR 的 T_g 向低温方向移动，两个 T_g 相互靠近。这可归因于 GO 在 XNBR 和 SBR 之间引入了相互作用，从而提高 XNBR 和 SBR 之间的相容性[80]。

综上所述，对于 GO 聚合物复合材料的诸多研究结果表明，GO 及还原得到的石墨烯应用于高分子复合材料中，具有提升其力学、电学、阻隔、热学等性能的应用优势。复合了 GO 的高分子复合材料已经被广泛地应用于超级电容器、医疗用品、耐高温型材料制造、阻隔薄膜以及耐低温型材料制造等研究领域，进一步提升了复合材料的性价比，甚至增添了新的功能，为石墨烯基复合材料的发展奠定了基础并提供了推动力[81]。

5.2 氧化石墨烯/无机非金属复合材料体系

除了在有机基体材料里作为功能材料添加，GO 和石墨烯也可与无机材料体系复合，发挥其性质并得到相关应用。

水泥基复合材料包括水泥浆、砂浆和混凝土，是最常用和最广泛使用的建筑材料。水泥基复合材料的主要缺点是脆性大、抗裂性差、抗拉强度低和应变能力差，这些缺点也是导致水泥基复合材料耐久性不良和维护成本高的主要原因[82]。目前，提高水泥基体强度和耐久性的传统方法主要有添加增强材料，使用化学矿物掺和料或降低水与胶凝材料的比例[84]。已有研究表明，纳米 SiO_2、纳米 Al_2O_3、碳纳米管、石墨烯、GO 等纳米增强材料能够与水泥基体构建新的相互作用机制，有效改善水泥复合材料的物理和化学性能[83~87]。

GO 纳米片表面存在亲水官能团，可以在水中形成稳定的悬浮液，对水泥基材料具有很高的亲和力，易于掺入水泥基材料中。关于 GO 改性水泥复合材料的研究已经很多。国内外相关研究表明，GO 对水泥基材料各项性能的影响非常显著，GO 的添加可以影响水泥基材料的水化过程，提升水泥基材料的力学性能和耐久性，GO 还可以用于水泥基复合材料的功能相，提高水泥基材料的吸附性能、电磁屏蔽性能、导电性能等[88~90]，因此在水泥复合材料中具有较好的应用前景[91]。

本节主要结合 GO 的结构特点，介绍 GO 在改善水泥基复合材料的力学性能、耐久性以及其他性能等方面的应用。

5.2.1 力学性能

如前所述，GO 的亲水性好，易于分散到水泥基复合材料中。表 5.3 总结了文献中 GO 对于水泥基复合材料力学性能的影响。由表 5.3 中的实验数据可见，添加 GO 能够提高水泥基复合材料早期和后期的力学强度。由于国内外各研究者所用的 GO 不同，所以实验结论中 GO 的最佳掺量以及对于水泥复合材料的提升效果也有较大差异。

表 5.3 GO 对水泥基复合材料力学性能的影响

基体	GO 添加量/%（质量）	水灰比	抗弯强度提升/（%/养护龄期）	抗压强度提升/（%/养护龄期）	参考文献
净浆	GO/0.05	0.29	90.5/28d	52.4/3d	[92]
净浆	GO/0.05	0.29	69.4/7d	66.4/7d	[93]
净浆	FGON/0.03	0.29	65.5/28d	51.3/28d	[94]
净浆	GO/0.03	0.29	55.0/28d	42.5/28d	[95]
净浆	GO/0.02	0.3	84.5/28d	60.1/28d	[96]
净浆	GO/0.04	0.3	43.2/28d	28.6/28d	[97]
净浆	GO/0.02	0.3	41.0/28d	13.0/28d	[98]
净浆	RGO/0.02	0.32	70.0/7d	22.0/28d	[99]
净浆	GO/0.03	0.33	56.6/27d	18.8/27d	[100]
净浆	GNPs/0.03	0.35	16.8/28d	1.3/28d	[101]
净浆	GO/0.04	0.4	—	15.1/28d	[102]
净浆	GO/0.05	0.4	16.2/15d	11.0/15d	[103]
净浆	GO/0.04	0.4	14.2/28d	37.0/28d	[104]
净浆	GO/0.022	0.42	26.7/3d	27.6/3d	[105]
净浆	GO/0.03	0.5	—	40.0/28d	[98]
净浆	GO/0.05	0.5	33.0/28d	59.0/28d	[106]
砂浆	GO/0.08	0.2	80.6/8d	—	[107]
砂浆	GO/0.03	0.37	70.7/3d	45.1/3d	[108]
砂浆	GO/0.05	0.37	106.4/14d	43.2/3d	[92]
砂浆	GO/0.022	0.42	34.1/7d	—	[109]
砂浆	GO/0.03	0.43	18.7/7d	—	[110]
砂浆	GO/1.0	0.45	—	114.1/14d	[111]
砂浆	GO/0.03	0.45	—	30.0/28d	[112]
砂浆	GO/0.125	0.45	—	53.0/3d	[113]
砂浆	GO/0.125	0.45	—	110.7/3d	[114]
砂浆	GO/0.02	0.45	36.7/3d	—	[115]
砂浆	FGON/0.1	0.48	70.8/15d	39.0/15d	[116]
砂浆	FGON/0.03	0.5	32.0/28d	20.3/28d	[117]
混凝土	GO/0.1	0.5	4.0/3d	14.2/7d	[118]

关于 GO 与水泥基复合材料的作用机制，研究者也有不同的观点，目前仍没有定论。水泥基复合材料本身是由水泥、水、砂、石等几种不同物质组合在一起形成的一种混合材料。所以，从宏观方面，其性能和组成材料有很大关系，水泥与水/胶凝材料的比例、GO 类型和养护龄期等因素对水泥基复合材料的机械强度都有很大影响。从微观方面，GO 的聚集、分散、尺寸和官能团也对水泥基复合材料的力学性能有影响。

目前，针对 GO 在提高水泥复合材料力学性能方面的作用机制主要有以下两个观点：a.GO 具有成核效应，由于 GO 片层的比表面积较高，可以作为水泥水化反应的成核生长点，有助于加速水泥复合材料的早期水化过程；GO 可以影响水泥水化产物的形态，形成更多的水化硅酸钙（C-S-H）；GO 纳米二维结构具有填充效应，可以改善水泥基体的孔隙结构，从而提高了水泥基复合材料的力学性能[104]；b.作为石墨烯的衍生物，GO 片层具有很高的面内拉伸强度，因此可以作为水泥复合材料中的增强填料，当水泥基复合材料受外力作用时，GO 片层能够有效地传递载荷，从而延缓水泥基体裂纹的产生，进而提高水泥基复合材料的力学性能[119]。

对于 GO 是否影响水泥水化过程这一问题，研究者也有不同的观点。有研究认为 GO 对水泥水化过程有影响。Hunain 等[120] 利用分子动力学的方法，模拟计算了石墨烯片层从 C-S-H 凝胶中拔出的界面强度，无官能团的石墨烯片层及单独带有羟基、硝基、羧基官能团的石墨烯片层的模拟强度分别为 1.2GPa、13.5GPa、6.1GPa、11.8GPa，这说明带有官能团的石墨烯片层与 C-S-H 凝胶的界面结合更好，更有利于对水泥基复合材料增强增韧。Lv 等[94,97,108] 人认为 GO 在水泥水化过程中可以起到模板作用，使水泥水化晶体产物排布更加规则，使硬化的水泥浆体的微观结构更加规整，进而提升水泥基材料的力学性能。吕升华等[97] 人还提出了 GO 对水泥水化晶体的调控机理：首先，GO 表面的含氧官能团优先与（硅酸三钙）C_3S，（硅酸二钙）C_2S 和（铝酸三钙）C_3A 反应并形成水化产物的生长点，水化产物的生长点和生长模式均由 GO 控制，称为模板效应。然后，GO 可以使同一GO 表面上的许多相邻的棒状水合晶体形成柱状和花形晶体。这些柱状产物由棒状的 AFt（三硫型水化硫酸铝钙）、AFm（单硫型水化硫酸铝钙）、CH（氢氧化钙）和 C-S-H（水化硅酸钙）组成，并且由于其周围的巨大应力而从 GO 表面沿相同方向向前生长，从而保持柱状，称为组装作用。最后，柱状晶体长成孔隙、裂缝或松散的结构，它们就会分开并形成完全开放的花状晶体，它们分散在孔隙和裂缝中作为填料和裂纹防止剂以延缓裂纹扩展。通常而言，在水泥复合材料的孔隙和间隙中形成的花状晶体和交联结构，对提高水泥复合材料的韧性有很大的贡献。

Lv 等[108] 人还研究了 GO 氧化程度对水泥基复合材料微观结构的影响。研究中对添加量为 0.03%（质量）的 GO（含氧量 25.43%（质量））水泥复合材料水化晶体在不同养护龄期时的产物进行了 SEM 表征，发现 GO 能够使水泥水化产物形成花朵状晶体，同时水泥基复合材料的拉伸强度、抗折强度和压缩强度比未添加

GO 的水泥基复合材料分别提高了 65.5%、60.7% 和 38.9%，表明 GO 可以起到增强增韧的作用。吕升华等还研究了 GO 的含氧量对于水泥基复合材料的影响，结果表明少量 GO 的掺入即可大大提高胶砂试件的机械强度，且其增强增韧效果较为显著。这一研究结果对于提高混凝土建筑物的抗裂性和耐久性、延长其使用寿命具有积极的意义。

Lin 等[121]人认为，Hummers' 法制备的比表面积为 $103m^2/g$ 的 GO 片层的含氧官能团在水化过程中可以提供给水和水泥颗粒更多的吸附位点，可作为晶核促进水泥水化产物的形成，同时 GO 表面还有一定的蓄水作用，为水泥颗粒的进一步水化提供水分子传输通道。这两方面的作用使得 GO 在水泥水化过程中可以起到催化作用，加速水泥的水化，从而进一步提升水泥石的各项性能。

还有研究者则认为 GO 对水泥水化进程并无影响，如 Horszczaruk 等[122]人通过拉曼光谱等方法发现，添加 GO 对于水泥水化动力学方面并无影响，当 GO 掺量为 3%（质量）时，对水泥基材料的杨氏模量具有显著增强作用。Zhu 等[106]人认为，GO 之所以能够提高水泥基材料的力学性能，主要归因于 GO 能够阻止水泥基体微裂缝的扩展，同时研究发现 GO 的添加量为 0.05%（质量）时，可使普通硅酸盐水泥的抗压强度提高 33%，弯曲强度提高 59%。对比水泥复合材料裂缝模式的微观形貌，普通水泥浆的裂缝通常以直通方式通过致密的水合产物，GO 水泥复合材料显示出了有分支和相当大的不连续性的许多细小的裂缝。这说明由于 GO 独特的二维结构，大比表面积可以有效地偏转裂纹，或者迫使裂纹在基体周围倾斜和扭曲。该过程可能有助于阻止细小裂纹，从而提升了水泥的抗折、抗压强度。

虽然众多研究发现 GO 对水泥水化产物及聚集状态具有调控作用，具有促使形成规整的水泥水化产物及聚集体结构的趋势，可使水泥基复合材料的力学性能有较大提高。但是，GO 在水泥复合材料中的应用仍然存在不容易均匀分散的问题，这也影响到水泥基复合材料的整体规整性及实验结果的重现性。

有学者研究了改性 GO 对于水泥复合材料的影响。曹明莉等[123]人将改性 GO 掺入水泥中，发现其对水泥水化晶体产物的形成有促进作用和模板效应，可以提高水泥基材料的强度和韧性。Abrishami 等[116]人利用 NH_2 作为 GO 的表面活化剂，与未经活化的 GO 相比，活化的 GO 更好地发挥增强水泥基复合材料强度的效果。Zhao 等[105]人采用聚羧酸系高效减水剂（PC）改性 GO，在一定温度下将两者混合得到 PC@GO，二者发生耦合反应，通过空间位阻和静电斥力作用对水泥颗粒起分散作用，从而提高 GO 在水泥基材料中的分散性能。

还有一些学者研究了 GO 与其他材料复配改性水泥基复合材料。Li 等[124]人研究了 GO 和 SWCNT 的复配对于水泥力学性能的影响。研究发现，GO 和碳纳米管具有协同效应，添加 1.5%（质量）的 GO 和 0.5%（质量）的 SWCNTs，能够使水泥复合材料的抗折强度提高 72.7%，高于二者单独添加的对于水泥复合材料强度的提高幅度（相比而言，单独 GO：51.2%；单独 SWCNTs：26.3%）。通过对硬化水泥石进行 XRD 测试发现，增加了水泥水化产物中 SiO_2 晶体的颗粒尺寸是 GO 和碳纳米管对水泥石抗折强度提高的主要原因。

吕生华等[125]人通过插层聚合反应制备了 GO 与丙烯酸（AA）和丙烯酰胺（AM）的复合物 GO/P（AA-AM），再将该复合物添加到水泥基复合材料中，可以解决 GO 纳米片层在水泥基材料中的团聚及难以均匀分散的问题。

也有学者研究了优化 GO 尺寸对于水泥复合材料性能的影响。杜涛等[126]人的研究表明，GO 片层的尺寸越大，对水泥净浆的力学性能和抗渗性能更有利，这是因为 GO 的尺寸越大，越有利于水泥水化产物 C-S-H 凝胶聚合，降低 $Ca(OH)_2$ 含量，起到增强增韧的效果。

5.2.2 耐久性

随着我国经济的发展以及对于基础建设的大力推进，高强、易施工、价廉的混凝土的用量日益增加，然而由于混凝土基体内部存在微裂缝和孔隙的缺陷，导致混凝土容易遭受一些腐蚀介质如氯盐、硫酸盐等的侵蚀，从而使混凝土构件的服役寿命缩短。利用纳米材料来提高混凝土结构的耐久性能已成为目前研究的重要内容。

Wang 等[92]人研究发现当 GO 的添加量为 0.02%（质量）时，可使水泥基复合材料的 28 天抗压和抗折强度分别提高 40.4% 和 90.5%，水泥基材料在 3 天龄期的放热量及放热速率下降 50%，这在很大程度上减少了由于水泥水化热的作用导致温度应力而出现裂缝。可见 GO 的添加既能够增强水泥基的力学强度，又能够减小外界腐蚀因子对水泥的侵蚀，从而提高了水泥的耐久性能。

Mohammed 等[112,127]人研究了 GO 水泥复合材料的抗氯离子侵蚀性能。实验结果表明，GO 的添加量为 0.01%（质量）时，水泥试块内部氯离子含量最少，这是因为 GO 改善了水泥水化产物的结构，减少了微观结构中的空隙，从而有效抑制氯离子的侵入，提高了水泥复合材料的耐久性。另外，添加 0.03%（质量）的 GO 后，水泥复合材料吸水率也显著提高，这说明 GO 的加入可以有效地提高水泥基体的传输性能，从而提高其耐久性。

Tong 等[128]人研究了 GO 对水泥复合材料的力学性能、抗侵蚀性能以及抗冻性能的影响。研究显示，GO 的掺入对混凝土和砂浆的力学性能有显著增强作用，能够有效提高水泥基材料的抗侵蚀能力。GO 则会提高砂浆的抗冻能力，经过 300 次冻融循环，纯砂浆的质量减少 0.96%，GO/水泥复合材料的质量减少 0.74%。

5.2.3 其他方面

曹明莉等[123]人认为 GO 二维表面上大量的含氧官能团可成为活性吸附位点吸附水溶液中的重金属离子，进而有效分离废水中的重金属离子。Chen 等[129]人在水泥中加入 0.04%（质量）的 GO 和碳纤维的复合物，在 X 波段（8.2~12.4GHz）

的吸收强度为34dB，在相同条件下，只加入碳纤维的吸收强度为26dB，相比提高了31%，可见GO增强了水泥的电磁屏蔽。Singh等[130]人的研究表明，添加GO和铁磁流体能够进一步提高水泥基复合材料的电磁屏蔽性能。当GO的添加量为30%（质量）时，GO/水泥复合材料在X波段可使材料的屏蔽效能达到46dB（即引起电磁波能量衰减99以上），这是因为水泥中的GO和铁磁流体的存在导致强烈的极化和磁矩，从而提高了磁损耗和屏蔽效能。Saafi等[131]人将GO与粉煤灰处理后得到的聚合物添加到水泥复合材料中，测试GO的含量分别为0%、0.1%和3.5%的试块，发现其电阻率随GO含量的增加而减小。

综上所述，作为一种新型纳米材料，GO可以明显地改善水泥基复合材料的机械性能，提高复合材料的抗渗透性；GO/水泥基复合材料的早期强度较高，可缩短施工周期；GO/水泥基复合材料具有超强的电磁屏蔽性能，可以减少影响人体健康的电磁辐射问题[132~134]。可见，GO在建筑材料中具有明显的应用优势，对于促进多功能建筑材料的进一步发展具有重要意义。

5.3 小结

氧化石墨烯具有众多特点，可以提高复合材料的各种性能。相比于传统纳米材料，除常规增强材料力学等性能外，氧化石墨烯的独特的片层结构使其在极少添加量的情况下即可提高复合材料屏蔽能力；而经过还原后，还可赋予复合材料导电、导热等独特功能。因此，氧化石墨烯对于发展新型功能型复合材料具有重要作用。

—— 参考文献 ——

[1] Coleman Jonathan N., Khan Umar, Blau Werner J., et al. Small but strong: A review of the mechanical properties of carbon nanotube-polymer composites [J]. Carbon, **2006**, 44（9）: 1624-1652.

[2] Costache Marius, Heidecker Matthew, Manias E, et al. The influence of carbon nanotubes, organically modified montmorillonites and layered double hydroxides on the thermal degradation and fire retardancy of polyethylene, ethylene-vinyl acetate copolymer and polystyrene [J]. Polymer, **2007**, 48（22）: 6532-6545.

[3] Homenick Christa, Lawson Gregor, Adronov Alex. Polymer Grafting of Carbon Nanotubes Using Living Free-Radical Polymerization [J]. Polymer Reviews, **2007**, 47（2）: 265-290.

[4] Liu Lei, Grunlan Jaime. Clay Assisted Dispersion of Carbon Nanotubes in Conductive Epoxy Nanocomposites [J]. Advanced Functional Materials, **2007**, 17（14）: 2343-2348.

[5] Kuilla Tapas, Bhadra Sambhu, Yao Dahu, et al. Recent advances in graphene based polymer composites [J]. Progress in Polymer Science, **2010**, 35（11）: 1350-1375.

[6] Tripathi Sandeep N., Rao G. S. Srinivasa, Mathur Ajit B., et al. Polyolefin/graphene nanocomposites: A review [J]. Rsc Advances, **2017**, 7（38）: 23615-23632.

[7] Dreyer Daniel R., Park Sungjin, Bielawski Christopher W., et al. The chemistry of graphene oxide. Chem Soc Rev [J]. Chemical Society Reviews, 2010, 39 (1): 228-240.

[8] Chen Yapeng, Gao Jingyao, Yan Qingwei, et al. Advances in graphene-based polymer composites with high thermal conductivity [J]. Veruscript Functional Nanomaterials, 2018, 2: OOSB06.

[9] Mohan Velram Balaji, Lau Kin Tak, Hui David, et al. Graphene-based materials and their composites: A review on production, applications and product limitations [J]. Composites Part B Engineering, 2018, 142: 200-220.

[10] Huang Yingjuan, Qin Yawei, Zhou Yong, et al. Polypropylene/Graphene Oxide Nanocomposites Prepared by In Situ Ziegler-Natta Polymerization [J]. Chemistry of Materials, 2010, 22 (13): 4096-4102.

[11] Gil Gonçalves, Paula A. A. P. Marques, Ana Barros-Timmons, Igor Bdkin, Manoj K. Singh, Nazanin Emami, José Grácio. Graphene oxide modified with PMMA via ATRP as a reinforcement filler [J]. Journal of Materials Chemistry, 2010, 20 (44): 9927-9934.

[12] Hwa Lee Sun, R Dreyer Daniel, An Jinho, et al. Polymer Brushes via Controlled, Surface-Initiated Atom Transfer Radical Polymerization (ATRP) from Graphene Oxide [J]. Macromolecular Rapid Communications, 2010, 31 (3): 281-288.

[13] Gao Tingting, Ye Qian, Pei Xiaowei, et al. Grafting Polymer Brushes on Graphene Oxide for Controlling Surface Charge States and Templated Synthesis of Metal Nanoparticles [J]. Journal of Applied Polymer Science, 2013, 127 (4): 3074-3083.

[14] Zhang Bin, Chen Yu, Xu Liqun, et al. Growing Poly (N-vinylcarbazole) from the Surface of Graphene Oxide via RAFT Polymerization [J]. Journal of Polymer Science Part A Polymer Chemistry, 2011, 49 (9): 2043-2050.

[15] Yu Dingshan, Yang Yan, Durstock Michael, et al. Soluble P3HT-grafted graphene for efficient bilayer-heterojunction photovoltaic devices [J]. Acs Nano, 2010, 4 (10): 5633-5640.

[16] Liu Yongsheng, Zhou Jiaoyan, Zhang Xiaoliang, et al. Synthesis, characterization and optical limiting property of covalently oligothiophene-functionalized graphene material [J]. Carbon, 2009, 47 (13): 3113-3121.

[17] Park Sungjin, Dikin Dmitriy A., Nguyen Son Binh T., et al. Graphene Oxide Sheets Chemically Cross-Linked by Polyallylamine [J]. Journal of Physical Chemistry C, 2009, 113 (36): 15801-15804.

[18] Wang Jen Yu, Yang Shin Yi, Huang Yuan Li, et al. Preparation and properties of graphene oxide/polyimide composite films with low dielectric constant and ultrahigh strength via in situ polymerization [J]. Journal of Materials Chemistry, 2011, 21 (35): 13569-13575.

[19] Xu Zhen, Gao Chao In situ Polymerization Approach to Graphene-Reinforced Nylon6 Composites [J]. Macromolecules, 2010, 43 (16): 6716-6723.

[20] Melucci Manuela, Treossi Emanuele, Ortolani Luca, et al. Facile covalent functionalization of graphene oxide using microwaves: Bottom-up development of functional graphitic materials [J]. Journal of Materials Chemistry, 2010, 20 (41): 9052-9060.

[21] Cao Yewen, Lai Zuliang, Feng Jiachun, et al. Graphene oxide sheets covalently functionalized with block copolymers via click chemistry as reinforcing fillers [J]. Journal of Materials Chemistry, 2011, 21 (25): 9271-9278.

[22] Sun Shengtong, Cao Yewen, Feng Jiachun, et al. Click chemistry as a route for the immobilization of well-defined polystyrene onto graphene sheets. [J]. Journal of Materials Chemistry, 2010, 20 (27): 5605-5607.

[23] Bao Chenlu, Guo Yuqiang, Song Lei, et al. In situ preparation of functionalized graphene oxide/epoxy

nanocomposites with effective reinforcements. J Mater Chem 21: 13290 [J]. Journal of Materials Chemistry, **2011**, 21 (35): 13290-13298.

[24] Guo Yuqiang, Bao Chenlu, Song Lei, et al. In Situ Polymerization of Graphene, Graphite Oxide, and Functionalized Graphite Oxide into Epoxy Resin and Comparison Study of On-the-Flame Behavior [J]. Ind. eng. chem. res, **2011**, 50 (13): 7772-7783.

[25] Huang Y. F., Lin C. W. Facile synthesis and morphology control of graphene oxide/polyaniline nanocomposites via in-situ polymerization process [J]. Polymer, **2012**, 53 (13): 2574-2582.

[26] Kim Hyunwoo, Miura Yutaka, Macosko Christopher W. Graphene/Polyurethane Nanocomposites for Improved Gas Barrier and Electrical Conductivity [J]. Chemistry of Materials, **2010**, 22 (11): 3441-3450.

[27] Li Yuqi, Pan Diyuan, Chen Shoubin, et al. In situ polymerization and mechanical, thermal properties of polyurethane/graphene oxide/epoxy nanocomposites [J]. Materials, **2013**, 47 (9): 850-856.

[28] Wang Xin, Hu Yuan, Song Lei, et al. In situ polymerization of graphene nanosheets and polyurethane with enhanced mechanical and thermal properties [J]. Journal of Materials Chemistry, **2011**, 21 (21): 4222-4227.

[29] Ferreira Filipe, Brito F. S., Franceschi Wesley, et al. Functionalized graphene oxide as reinforcement in epoxy based nanocomposites [J]. Surfaces Interfaces, **2018**, 10: 100-109.

[30] Strankowski Michał, Korzeniewski Piotr, Strankowska Justyna, et al. Morphology, Mechanical and Thermal Properties of Thermoplastic Polyurethane Containing Reduced Graphene Oxide and Graphene Nanoplatelets [J]. Materials, **2018**, 11 (1): 82.

[31] Potts Jeffrey R., Sun Hwa Lee, Alam Todd M., et al. Thermomechanical properties of chemically modified graphene/poly (methyl methacrylate) composites made by in situ polymerization [J]. Carbon, **2011**, 49 (8): 2615-2623.

[32] Tripathi Sandeep Nath, Saini Parveen, Gupta Deeksha, et al. Electrical and mechanical properties of PMMA/reduced graphene oxide nanocomposites prepared via in situ polymerization [J]. Journal of Materials Science, **2013**, 48 (18): 6223-6232.

[33] Kan Lanyan, Xu Zhen, Gao Chao. General Avenue to Individually Dispersed Graphene Oxide-Based Two-Dimensional Molecular Brushes by Free Radical Polymerization [J]. Macromolecules, **2011**, 44 (3): 444-452.

[34] Wang Xing, Li Hengyi, Huang Guangsu, et al. Graphene oxide induced crosslinking and reinforcement of elastomers [J]. Composites Science and Technology, **2017**, 144: 223-229.

[35] Tsagkalias I., Manios T., Achilias D.. Effect of Graphene Oxide on the Reaction Kinetics of Methyl Methacrylate In Situ Radical Polymerization via the Bulk or Solution Technique [J]. Polymers, **2017**, 9 (9): 432.

[36] Layek Rama K. and Nandi Arun K. A review on synthesis and properties of polymer functionalized graphene [J]. Polymer, **2013**, 54 (19): 5087-5103.

[37] 杨永岗, 陈成猛, 温月芳, 等. 氧化石墨烯及其与聚合物的复合 [J]. 新型炭材料, **2008**, 23 (3): 193-200.

[38] 赵茜, 邱东方, 王晓燕, 刘天西. 壳聚糖/氧化石墨烯纳米复合材料的形态和力学性能研究 [J]. 化学学报, **2011**, 69 (10): 1259-1263.

[39] Wang Yan, Shi Zixing, Fang Jianhua, et al. Graphene oxide/polybenzimidazole composites fabricated by a solvent-exchange method [J]. Carbon, **2011**, 49 (4): 1199-1207.

[40] Liang Jiajie, Huang Yi, Zhang Long, et al. Molecular-Level Dispersion of Graphene into Poly (vinyl al-

cohol) and Effective Reinforcement of their Nanocomposites [J]. Advanced Functional Materials, **2009**, 19 (14): 2297-2302.

[41] Liu Pengzhang, Zhang Xumin, Jia Hongbing, et al. High mechanical properties, thermal conductivity and solvent resistance in graphene oxide/styrene-butadiene rubber nanocomposites by engineering carboxylated acrylonitrile-butadiene rubber [J]. Composites Part B Engineering, **2017**, 130: 257-266.

[42] Liu Xuan, Kuang Wenyi, Guo Baochun. Preparation of rubber/graphene oxide composites with in-situ interfacial design [J]. Polymer, **2015**, 56: 553-562.

[43] Yin Biao, Wang Jingyi, Jia Hongbing, et al. Enhanced mechanical properties and thermal conductivity of styrene-butadiene rubber reinforced with polyvinylpyrrolidone-modified graphene oxide [J]. Journal of Materials Science, **2016**, 51 (12): 5724-5737.

[44] Wu Jinrong, Huang Guangsu, Li Hui, et al. Enhanced mechanical and gas barrier properties of rubber nanocomposites with surface functionalized graphene oxide at low content [J]. Polymer, **2013**, 54 (7): 1930-1937.

[45] Tang Zhenghai, Wu Xiaohui, Guo Baochun, et al. Preparation of butadiene-styrene-vinyl pyridine rubber-graphene oxide hybrids through co-coagulation process and in situ interface tailoring [J]. Journal of Materials Chemistry, **2012**, 22 (15): 7492-7501.

[46] Valentini L., Bon S. Bittolo, Hernández M., et al. Nitrile butadiene rubber composites reinforced with reduced graphene oxide and carbon nanotubes show superior mechanical, electrical and icephobic properties [J]. Composites Science Technology, **2018**, 166: 109-114.

[47] Wang Jianfeng, Jin Xiuxiu, Zhang Xiaomeng, et al. Achieving high-performance poly (styrene-b-ethylene-ranbutylene-b-styrene) nanocomposites with tannic acid functionalized graphene oxide [J]. Composites Science Technology, **2018**, 158: 137-146.

[48] 邓尧, 黄肖容, 邬晓龄. 氧化石墨烯复合材料的研究进展 [J]. 材料导报, **2012**, 26 (15): 84-87.

[49] King Julia A., Via Michael D., King Michelle E., et al. Electrical and Thermal Conductivity and Tensile and Flexural Properties: Comparison of Carbon Black/Polycarbonate and Carbon Nanotube/Polycarbonate Resins [J]. Journal of Applied Polymer Science, **2011**, 121 (4): 2273-2281.

[50] Hu Huating, Wang Xianbao, Wang Jingchao, et al. Preparation and properties of graphene nanosheets-polystyrene nanocomposites via in situ emulsion polymerization [J]. Chemical Physics Letters, **2010**, 484 (4): 247-253.

[51] Zhang Hao Bin, Zheng Wen Ge, Yan Qing, et al. Electrically conductive polyethylene terephthalate/graphene nanocomposites prepared by melt compounding [J]. Polymer, **2010**, 51 (5): 1191-1196.

[52] Liu Na, Luo Fang, Wu Haoxi, et al. One-Step Ionic-Liquid-Assisted Electrochemical Synthesis of Ionic-Liquid-Functionalized Graphene Sheets Directly from Graphite [J]. Advanced Functional Materials, **2008**, 18 (10): 1518-1525.

[53] 程博, 齐暑华, 何栋, 等. 纳米石墨微片/聚氯乙烯复合材料的制备与性能 [J]. 复合材料学报, **2012**, 29 (1): 8-15.

[54] Li Hui, Wu Siduo, Wu Jinrong, et al. Enhanced electrical conductivity and mechanical property of SBS/graphene nanocomposite [J]. Journal of Polymer Research, **2014**, 21 (5): 456.

[55] 陈翔峰, 陈国华, 吴大军, 等. AS/石墨纳米薄片复合材料的制备 [J]. 功能材料, **2004**, 35 (z1): 1007-1008.

[56] 吕生华, 李莹, 杨文强, 等. 氧化石墨烯/壳聚糖生物复合材料的制备及应用研究进展 [J]. 材料工程, **2016**, 44 (10): 119-128.

[57] Gao Yang. Graphene and Polymer Composites for Supercapacitor Applications: a Review [J]. Nanoscale Research Letters, **2017**, 12 (1): 387.

[58] Wang Xinyu, Wan Fang, Zhang Linlin, et al. Large-Area Reduced Graphene Oxide Composite Films for Flexible Asymmetric Sandwich and Microsized Supercapacitors [J]. Advanced Functional Materials, **2018**, 28 (18): 1707247.

[59] Kashiwagi Takashi, Du Fangming, Douglas J, et al. Nanoparticle networks reduce the flammability of polymer nanocomposites [J]. Nature Materials, **2005**, 4 (12): 928-933.

[60] Kashiwagi Takashi, Mu Minfang, Winey Karen, et al. Relation between the viscoelastic and flammability properties of polymer nanocomposites [J]. Polymer, **2008**, 49 (20): 4358-4368.

[61] Xiong Xiaogang, Wang Jingyi, Jia Hongbing, et al. Structure, thermal conductivity, and thermal stability of bromobutyl rubber nanocomposites with ionic liquid modified graphene oxide [J]. Polymer Degradation Stability, **2013**, 98 (11): 2208-2214.

[62] Wang Huishan, Qiu Zhaobin. Crystallization behaviors of biodegradable poly (l-lactic acid) /graphene oxide nanocomposites from the amorphous state [J]. Thermochimica acta, **2011**, 526 (1-2): 229-236.

[63] 李芸博. 氧化石墨烯复合材料的研究进展分析 [J]. 化工管理, **2017**, (15): 109-110.

[64] 钟涛, 杨娟, 周亚洲, 等. 纳米银-氧化石墨烯复合材料抗菌性能研究进展 [J]. 材料导报, **2014**, (s1): 64-66.

[65] 吕生华, 朱琳琳, 李莹, 等. 氧化石墨烯复合材料的研究现状及进展 [J]. 材料工程, **2016**, 44 (12): 107-117.

[66] Veca L. Monica, Meziani Mohammed J., Wang Wei, et al. Carbon Nanosheets for Polymeric Nanocomposites with High Thermal Conductivity [J]. Advanced Materials, **2010**, 21 (21): 2088-2092.

[67] Yu Aiping, Ramesh Palanisamy, Itkis M. E., et al. Graphite Nanoplatelet-Epoxy Composite Thermal Interface Materials [J]. Journal of Physical Chemistry C, **2007**, 111 (21): 7565-7569.

[68] Ganguli Sabyasachi, Roy Ajit K., and Anderson David P. Improved thermal conductivity for chemically functionalized exfoliated graphite/epoxy composites [J]. Carbon, **2008**, 46 (5): 806-817.

[69] Mu Qiuhong, Feng Shengyu, Diao Guangzhao. Thermal conductivity of silicone rubber filled with ZnO [J]. Polymer Composites, **2007**, 28 (2): 125-130.

[70] Zhao Xiongwei, Zang Chongguang, Wen Yuquan, et al. Thermal and mechanical properties of liquid silicone rubber composites filled with functionalized graphene oxide [J]. Journal of Applied Polymer Science, **2015**, 132 (38).

[71] Thomas Shaji P., Thomas Saliney, Mathew E. J., et al. Transport and electrical properties of natural rubber/nitrile rubber blend composites reinforced with multiwalled carbon nanotube and modified nano zinc oxide [J]. Polymer Composites, **2014**, 35 (5): 956-963.

[72] Cao Yewen, Zhang Jing, Feng Jiachun, et al. Compatibilization of immiscible polymer blends using graphene oxide sheets [J]. Acs Nano, **2011**, 5 (7): 5920-5927.

[73] Biswas T, Debnath Subhas, Naskar N, et al. SBR-XNBR blends: a novel approach towards compatibilization [J]. European Polymer Journal, **2004**, 40 (4): 847-854.

[74] Yang Jinghui, Feng Chenxia, Dai Jian, et al. Compatibilization of immiscible nylon 6/poly (vinylidene fluoride) blends using graphene oxides [J]. Polymer International, **2013**, 62 (7): 1085-1093.

[75] Kang Hailan, Zuo Kanghua, Wang Zhao, et al. Using a green method to develop graphene oxide/elastomers nanocomposites with combination of high barrier and mechanical performance [J]. Composites Science Technology, **2014**, 92 (3): 1-8.

[76] Laskowska Anna, Marzec Anna, Zaborski Marian, et al. Reinforcement of carboxylated acrylonitrile-bu-

tadiene rubber (XNBR) with graphene nanoplatelets with varying surface area [J]. Journal of Polymer Engineering, **2014**, 34 (9): 883-893.

[77] Yan Ning, Xia Hesheng, Wu Jinkui, et al. Compatibilization of natural rubber/high density polyethylene thermoplastic vulcanizate with graphene oxide through ultrasonically assisted latex mixing [J]. Journal of applied polymer science, **2013**, 127 (2): 933-941.

[78] Chen Biyan, Ma Nan, Bai Xin, et al. Effects of graphene oxide on surface energy, mechanical, damping and thermal properties of ethylene-propylene-diene rubber/petroleum resin blends [J]. Rsc Advances, **2012**, 2 (11): 4683-4689.

[79] Samaržija-Jovanović Suzana, Jovanović Vojislav, Marković Gordana, et al. Nanocomposites based on silica-reinforced ethylene-propylene-diene-monomer/acrylonitrile-butadiene rubber blends [J]. Composites Part B: Engineering, **2011**, 42 (5): 1244-1250.

[80] Razak Jeefferie Abd, Ahmad Sahrim Haji, Ratnam Chantara Thevy, et al. Effects of poly (ethyleneimine) adsorption on graphene nanoplatelets to the properties of NR/EPDM rubber blend nanocomposites [J]. Journal of Materials Science, **2015**, 50 (19): 6365-6381.

[81] 张文文, 刘秀军, 李同起, 等. 氧化石墨烯/聚合物复合材料的研究进展 [J]. 化工新型材料, **2015**, 43 (1): 12-14.

[82] Sagar R. Vidya, Prasad B. K. Raghu, Kumar S. Shantha. An experimental study on cracking evolution in concrete and cement mortar by the b -value analysis of acoustic emission technique [J]. Cement Concrete Research, **2012**, 42 (8): 1094-1104.

[83] Muthu Murugan, Santhanam Manu. Effect of reduced graphene oxide, alumina and silica nanoparticles on the deterioration characteristics of Portland cement paste exposed to acidic environment [J]. Cement Concrete Composites, **2018**, 91: 118-137.

[84] Zhang Rui, Cheng Xin, Hou Pengkun, et al. Influences of nano-TiO_2 on the properties of cement-based materials: Hydration and drying shrinkage [J]. Construction Building Materials, **2015**, 81: 35-41.

[85] Hakamy A., Shaikh F. U. A., Low I. M.. Effect of calcined nanoclay on the durability of NaOH treated hemp fabric-reinforced cement nanocomposites [J]. Materials Design, **2016**, 92: 659-666.

[86] Singh N. B., Kalra Meenu, Saxena S. K.. Nanoscience of Cement and Concrete [J]. Materials Today: Proceedings, **2017**, 4 (4): 5478-5487.

[87] Bastos Guillermo, Patiño F, Cambeiro Faustino, et al. Nano-Inclusions Applied in Cement-Matrix Composites: A Review [J]. Materials, **2016**, 9 (12): 1015.

[88] Xu Yidong, Zeng Juqing, Chen Wei, et al. A holistic review of cement composites reinforced with graphene oxide [J]. Construction and Building Materials, **2018**, 171: 291-302.

[89] Li Xiangyu, Wang Linhao, Liu Yuqing, et al. Dispersion of graphene oxide agglomerates in cement paste and its effects on electrical resistivity and flexural strength [J]. Cement Concrete Composites, **2018**, 92: 145-154.

[90] Yang Haibin, Monasterio Manuel, Cui Hongzhi, et al. Experimental study of the effects of graphene oxide on microstructure and properties of cement paste composite [J]. Composites Part A: Applied Science and Manufacturing, **2017**, 102: 263-272.

[91] Ghazizadeh Sam, Duffour Philippe, Skipper Neal T., et al. Understanding the behaviour of graphene oxide in Portland cement paste [J]. Cement Concrete Research, **2018**, 111: 169-182.

[92] Wang Qin, Wang Jian, Lu Chun Xiang, et al. Influence of graphene oxide additions on the microstructure and mechanical strength of cement [J]. New Carbon Materials, **2015**, 30 (4): 349-356.

[93] Lv Shenghua, Ting Sun, Liu Jingjing, et al. Use of graphene oxide nanosheets to regulate the microstructure of hardened cement paste to increase its strength and toughness [J]. CrystEngComm, 2014, 16 (36): 8508-8516.

[94] Lv SH, Deng LJ, Yang WQ, et al. Fabrication of polycarboxylate/graphene oxide nanosheet composites by copolymerization for reinforcing and toughening cement composites [J]. Cement Concrete Composites, 2016, 66: 1-9.

[95] Wang Qin, Cui Xinyou, Wang Jian, et al. Effect of fly ash on rheological properties of graphene oxide cement paste [J]. Construction Building Materials, 2017, 138: 35-44.

[96] Lv Shenghua, Ma Yujuan, Qiu Chaochao, et al. Regulation of GO on cement hydration crystals and its toughening effect [J]. Magazine of Concrete Research, 2013, 65 (20): 1246-1254.

[97] Lv Shenghua, Liu Jingjing, Sun Ting, et al. Effect of GO nanosheets on shapes of cement hydration crystals and their formation process [J]. Construction Building Materials, 2014, 64 (64): 231-239.

[98] Gong Kai, Pan Zhu, Korayem Asghar H, et al. Reinforcing effects of graphene oxide on portland cement paste [J]. Journal of Materials in Civil Engineering, 2015, 27 (2): 1-6.

[99] Murugan M., Santhanam Manu, Gupta Soujit Sen, et al. Influence of 2D rGO nanosheets on the properties of OPC paste [J]. Cement Concrete Composites, 2016, 70: 48-59.

[100] Wang Min, Wang Rumin, Yao Hao, et al. Study on the three dimensional mechanism of graphene oxide nanosheets modified cement [J]. Construction Building Materials, 2016, 126: 730-739.

[101] Wang Baomin, Jiang Ruishuang, Wu Zhenlin. Investigation of the Mechanical Properties and Microstructure of Graphene Nanoplatelet-Cement Composite [J]. Nanomaterials, 2016, 6 (11): 200.

[102] Shang Yu, Zhang Dong, Yang Chao, et al. Effect of graphene oxide on the rheological properties of cement pastes [J]. Construction Building Materials, 2015, 96: 20-28.

[103] Lu Zeyu, Hou Dongshuai, Meng Lingshi, et al. Mechanism of cement paste reinforced by graphene oxide/carbon nanotubes composites with enhanced mechanical properties [J]. Rsc Advances, 2015, 5 (122): 100598-100605.

[104] Li Wengui, Li Xiangyu, Chen Shu Jian, et al. Effects of graphene oxide on early-age hydration and electrical resistivity of Portland cement paste [J]. Construction Building Materials, 2017, 136: 506-514.

[105] Zhao Li, Guo Xinli, Ge Chuang, et al. Investigation of the effectiveness of PC@GO on the reinforcement for cement composites [J]. Construction Building Materials, 2016, 113: 470-478.

[106] Zhu Pan, He Li, Qiu Ling, et al. Mechanical properties and microstructure of a graphene oxide-cement composite [J]. Cement Concrete Composites, 2015, 58: 140-147.

[107] Lu Cong, Lu Zeyu, Li Zongjin, et al. Effect of graphene oxide on the mechanical behavior of strain hardening cementitious composites [J]. Construction Building Materials, 2016, 120: 457-464.

[108] Lv Shenghua, Ma Yujuan, Qiu Chaochao, et al. Effect of graphene oxide nanosheets of microstructure and mechanical properties of cement composites [J]. Construction Building Materials, 2013, 49 (12): 121-127.

[109] Zhao Li, Guo Xinli, Ge Chuang, et al. Mechanical behavior and toughening mechanism of polycarboxylate superplasticizer modified graphene oxide reinforced cement composites [J]. Composites Part B: Engineering, 2017, 113: 308-316.

[110] Qian Ye, Abdallah Maika Yzabelle, Kawashima Shiho. Characterization of cement-based materials modified with graphene-oxide [J]. Nanotechnology in Construction, 2015, 259-264.

[111] Sharma Snigdha, Kothiyal N. C.. Influence of graphene oxide as dispersed phase in cement mortar matrix in defining the crystal patterns of cement hydrates and its effect on mechanical, microstructural and crys-

tallization properties [J]. RSC Advances, **2015**, 5 (65): 52642-52657.

[112] Mohammed A, Sanjayan JG, Duan WH, et al. Graphene oxide impact on hardened cement expressed in enhanced freeze-thaw resistance [J]. Journal of Materials in Civil Engineering, **2016**, 28 (9): 04016072.

[113] Sharma Snigdha, Kothiyal N. C.. Comparative effects of pristine and ball-milled graphene oxide on physico-chemical characteristics of cement mortar nanocomposites [J]. Construction Building Materials, **2016**, 115: 256-268.

[114] Kothiyal N. C., Sharma Snigdha, Mahajan Swati, et al. Characterization of reactive graphene oxide synthesized from ball-milled graphite: its enhanced reinforcing effects on cement nanocomposites [J]. Journal of Adhesion Science Technology, **2016**, 30 (9): 915-933.

[115] Zhao Li, Guo Xinli, Liu Yuanyuan, et al. Synergistic effects of silica nanoparticles/polycarboxylate superplasticizer modified graphene oxide on mechanical behavior and hydration process of cement composites [J]. Rsc Advances, **2017**, 7 (27): 16688-16702.

[116] Abrishami M Ebrahimizadeh, Zahabi V. Reinforcing graphene oxide/cement composite with NH_2 functionalizing group [J]. Bulletin of Materials Science, **2016**, 39 (4): 1073-1078.

[117] Cao Mingli, Zhang Huixia, Zhang Cong. Effect of graphene on mechanical properties of cement mortars [J]. Journal of Central South University, **2016**, 23 (4): 919-925.

[118] Devasena M, Karthikeyan J. Investigation on strength properties of graphene oxide concrete [J]. International Journal of Engineering Science Invention Research Development, **2015**, 1 (8): 307-310.

[119] Ghazizadeh S, Duffour P, Skipper NT, et al. An investigation into the colloidal stability of graphene oxide nano-layers in alite paste [J]. Cement Concrete Research, **2017**, 99: 116-128.

[120] Alkhateb Hunain, Al-Ostaz Ahmed, Cheng Alexander H-D, et al. Materials genome for graphene-cement nanocomposites [J]. Journal of Nanomechanics Micromechanics, **2013**, 3 (3): 67-77.

[121] Lin Changqing, Wei Wei, Hu Yunhang. Catalytic behavior of graphene oxide for cement hydration process [J]. Journal of Physics Chemistry of Solids, **2016**, 89 (3): 128-133.

[122] Horszczaruk Elżbieta, Mijowska Ewa, Kalenczuk Ryszard, et al. Nanocomposite of cement/graphene oxide-Impact on hydration kinetics and Young's modulus [J]. Construction building Materials, **2015**, 78: 234-242.

[123] 曹明莉, 张会霞, 张聪. 石墨烯对水泥净浆力学性能及微观结构的影响 [J]. 哈尔滨工业大学学报, **2015**, 47 (12): 26-30.

[124] Li Xueguang, Wei Wei, Qin Hao, et al. Co-effects of graphene oxide sheets and single wall carbon nanotubes on mechanical properties of cement [J]. Chemistry of Solids, **2015**, 85: 39-43.

[125] 吕生华, 张佳, 朱琳琳, 等. 氧化石墨烯对水泥基复合材料微观结构的调控作用及对抗压抗折强度的影响 [J]. 化工学报, **2017**, 68 (6): 2585-2595.

[126] 杜涛. 氧化石墨烯水泥基复合材料性能研究. 哈尔滨: 哈尔滨工业大学. 2014.

[127] Mohammed A, Sanjayan Jay, Duan W. H., et al. Incorporating graphene oxide in cement composites: A study of transport properties [J]. Construction Building Materials, **2015**, 84: 341-347.

[128] Tong Teng, Fan Zhou, Liu Qiong, et al. Investigation of the effects of graphene and graphene oxide nanoplatelets on the micro- and macro-properties of cementitious materials [J]. Construction Building Materials, **2016**, 106: 102-114.

[129] Chen Juan, Zhao Dan, Ge Heyi, et al. Graphene oxide-deposited carbon fiber/cement composites for electromagnetic interference shielding application [J]. Construction Building Materials, **2015**, 84: 66-72.

[130] Singh Avanish Pratap,Monika Mishra,Amita Chandra,et al. Graphene oxide/ferrofluid/cement composites for electromagnetic interference shielding application [J]. Nanotechnology,**2011**,22 (46): 465701.

[131] Saafi Mohamed,Tang Leung,Fung Jason,et al. Graphene/fly ash geopolymeric composites as self-sensing structural materials [J]. Smart Materials Structures,**2014**,23 (6): 065006.

[132] Long Wu-Jian,Wei Jing-Jie,Xing Feng,et al. Enhanced dynamic mechanical properties of cement paste modified with graphene oxide nanosheets and its reinforcing mechanism [J]. Cement Concrete Composites,**2018**,93: 127-139.

[133] Long Wu Jian,Zheng Dan,Duan Hua Bo,et al. Performance enhancement and environmental impact of cement composites containing graphene oxide with recycled fine aggregates [J]. Journal of Cleaner Production,**2018**,194: 193-202.

[134] Roy Rahul,Mitra Ananda,Ganesh Ajay T,et al. Effect of Graphene Oxide Nanosheets dispersion in cement mortar composites incorporating Metakaolin and Silica Fume [J]. Construction Building Materials,**2018**,186: 514-524.

第6章
CHAPTER 6

氧化石墨烯在能量存储中的应用

随着人类对能源与日俱增的需求，寻找清洁能源及其存储方式是当代科学的研究发展方向。石墨烯作为一种二维碳材料，凭借其独特的物理化学性质，在新能源研究及实际应用中得到了广泛的关注，为能源领域的不断发展提供了无限潜力。GO 作为石墨烯的一种衍生物，其中大量的含氧官能团使其成为石墨烯功能化应用的重要物质，GO 及其复合物在锂离子电池、超级电容器、燃料电池、太阳能电池等领域有了越来越多的发展和应用，促进了新能源领域的快速进步，对提高能源的利用效率、节能减排及环境保护具有重要意义。本章主要介绍石墨烯在锂离子电池电极材料、超级电容器材料、燃料电池材料和太阳能电池材料中的研究进展。

6.1 氧化石墨烯在锂离子电池中应用的研究进展

20 世纪 90 年代，日本索尼公司提出了以碳材料为负极、含锂的化合物为正极的电池，也就是锂离子电池。为了提高锂离子电池的性能，人们研究过很多纳米材料，包括石墨烯及其衍生物，其中以 GO 为原料制备得到的石墨烯具有高导电性、大比表面积等优良特性，使其成为当前锂离子电池领域的研究热点[1]。石墨烯在锂离子电池中的应用主要包括以下几种途径：石墨烯直接用作锂离子电池负极材料、石墨烯与负极材料的复合、石墨烯与正极材料的复合、石墨烯作为正极材料中的导电剂。

6.1.1 氧化石墨烯用作锂离子电池负极材料

目前锂离子电池的负极材料以石墨为主,现阶段几乎达到其理论容量值,因此高容量负极材料成为了当前锂离子电池中的研究热点。负极材料,应该具有良好的锂离子吸附和电子传输能力。石墨烯表面可以存储锂离子,具有高的电子迁移能力。石墨烯作为负极材料还可以缩短锂离子的传输路径。Bulusheva 等[2]人将 GO 置于浓硫酸中加热,之后在惰性气体中进行高温煅烧得到表面有 2~5nm 孔的石墨烯,该石墨烯材料具有良好的倍率性能。华南理工大学的 Lian 等[3] 人将 GO 置于高温煅烧炉中,在惰性气体的保护下还原得到层数少、缺陷少、杂质少的高质量石墨烯,并将其用作锂离子电池负极材料。该负极材料在 100mA/g 的电流密度下其充放电容量高达 1264mA·h/g;在循环 40 次后其容量仍可以达到 848mA·h/g;当电流密度增大到 1000mA/g 充放电时,其容量还可保持在 420mA·h/g。将 GO 通过各种方法处理得到的石墨烯作为负极可以大大提高电池的容量,但是其首次库仑效率较低,电池的循环寿命缩短,还可能降低锂离子电池负极的压实密度,这样会导致电池的能量密度大大下降,难以直接作为负极使用。

6.1.2 氧化石墨烯与负极材料的复合在锂离子电池中的应用

目前的负极材料中,硅被认为是最具有潜力的负极材料之一,因为它在自然界中含量多,还具有低的嵌锂电位和很高的理论比容量。然而,在锂离子脱嵌过程中,硅的体积变化比较明显,使得材料与负极集流体之间黏结性变差,造成电池循环性能的大幅度下降。同时硅还会在电池循环过程中出现团聚现象,引起电池容量的迅速下降。将硅材料和石墨烯进行复合得到复合负极材料,其中石墨烯可以抑制硅材料在充放电过程中的团聚,减缓硅材料的体积变化,从而提高电池的容量和循环性能。此外,石墨烯还有助于电解液的浸润,从而提高电池的性能。He 等[4]人通过喷雾干燥法制备了一种高性能的石墨烯/硅复合材料。研究中首先将 GO 与纳米硅超声混合,通过喷雾干燥后在 700℃下进行煅烧得到复合材料,在 200mA/g 的电流密度下充放电 30 次后,容量仍可达到 1502mA·h/g,其容量保持率为 98%,具有良好的循环性能。Lee 等[5] 人报道了三维结构的硅/石墨烯复合材料。在制备过程中,将 GO 与硅进行复合得到 GO/硅的复合纸,接着在 700℃条件下高温还原得到硅/石墨烯复合材料,该材料作为锂离子负极材料,在循环 200 次后容量仍高达 1500mA·h/g,表现出了优异的循环性能。Li 等[6] 人制备了壳聚糖和 GO 双层包覆 Si 颗粒复合材料,由于 Si 颗粒表面双包覆层的协同作用,材料有优良的循环性能,复合材料在 100 个循环后具有 935.77mA·h/g 和 71.9% 的剩余容量。

过渡金属氧化物也成为另一种可选择的锂电池负极材料。这类材料可以具有高的理论容量,但是导电性通常较差。有的氧化物和硅材料类似,在充放电过程中有很明显的体积变化,从而导致电池的循环问题。通过将石墨烯与这些金属氧化物进行复合改性，可改善电池的循环性能。文献已经报道了多种金属氧化物与石墨烯进行复合（Fe_2O_3、Fe_3O_4、TiO_2、Co_3O_4、Mn_3O_4等）的研究结果[7]。例如，虽然Co_3O_4的理论容量是石墨理论容量的2.4倍，但是其在循环过程中容量衰减很快，这是由于锂离子在嵌入/脱出过程中引起Co_3O_4的团聚和明显的体积变化。Wu等[8]人制备了石墨烯/Co_3O_4复合材料。其制备过程是将GO进行热处理膨胀后接着在高温下还原得到石墨烯，再加入钴盐后合成得到石墨烯/CO_3O_4复合材料。该复合材料在30次的充放电循环后，其容量还可以达到935mA·h/g。Hao等[9]人制备得到了三维结构的Fe_3O_4/石墨烯复合材料。制备过程也是在GO中加入铁盐后高温处理得到。由于材料中石墨烯对充放电过程中Fe_3O_4体积的抑制作用，使得其具有很好的电化学循环性能，在1C充放电循环500圈后仍有1000mA·h/g的容量保留。

6.1.3 氧化石墨烯与正极材料的复合在锂离子电池中的应用

高能量密度的锂离子电池正是现阶段新能源追求发展的目标，而正极材料的能量密度是目前锂离子电池进一步提升能量密度的主要瓶颈之一。目前主流的正极材料有$LiCoO_2$、$LiFePO_4$、$LiMn_2O_4$和$LiNi_{1/3}Co_{1/3}Mn_{1/3}O_2$等。其中$LiFePO_4$由于其高安全性和长循环寿命，是目前广泛使用的商业动力锂离子正极材料，但是$LiFePO_4$较低的电子电导率和离子扩散率，限制了其倍率性能。将石墨烯与$LiFePO_4$正极材料进行复合可以改善其倍率性能。Wang等[10]人通过在GO溶液中加入铁盐、锂盐等后水热处理，得到$LiFePO_4$/石墨烯复合材料并用于锂离子电池正极，在电压范围2.0~4.2V、0.1C放电倍率下容量为160.3mA·h/g，当放电倍率增大到10C时其容量仍可保持在81.5mA·h/g，这均比单独的$LiFePO_4$有所提高。Ding等[11]人采用共沉淀法制备得到$LiFePO_4$/石墨烯复合材料。制备过程中通过水合肼将GO进行化学还原，接着再与铁盐等材料反应得到该复合材料，在电压范围2.4~4.2V、10C放电倍率下其容量还有110mA·h/g。Wang等[12]人通过溶胶-凝胶法制备得到N-掺杂的石墨烯与$LiFePO_4$@C纳米颗粒复合材料。该材料是先将石墨进行氧化后进行氮掺杂制备得到氮掺杂GO，之后在氮掺杂GO表面合成前驱体并进行高温煅烧得到复合材料。该研究中石墨烯的加入也大大提高了$LiFePO_4$正极材料的性能，在10C大倍率放电下容量仍可保留118.6mA·h/g。性能提升的原因一方面是由于石墨烯的高电子电导率，另一方面是由于复合材料中会有大量的孔结构，从而使电解液更快的扩散到活性材料，使其表现出更好的电解液浸润性和更高的倍率性能。石墨烯包覆$LiFePO_4$后，还有可能减少电极与电解液之间发生的副反应，有助于形成更稳定的固体电解质界面膜（SEI膜）。随着能量密度更高的要求，三元材料例如$LiNi_{1/3}Co_{1/3}Mn_{1/3}O_2$，由于其高理论容量，近几年来引起了新能源行业的关注，但是其电子电导率较低，首次库伦效率也低。Rao

等[13]人使用微乳液法和球磨法制备得到 $LiNi_{1/3}Co_{1/3}Mn_{1/3}O_2$ 和石墨烯的复合材料，利用石墨烯的高电子电导率提高其电池性能，其中石墨烯材料也是通过将 GO 进行还原得到。复合材料在 2.5～4.4V 电压范围内进行充放电，当放电倍率为 0.05C 时，其容量为 185mA·h/g，增加放电倍率到 5C 时其容量仍可保持在 153mA·h/g。在上述制备中，基本都是在制备 $LiFePO_4$ 和 $LiNi_{1/3}Co_{1/3}Mn_{1/3}O_2$ 的过程中原位复合 GO，在后续的处理过程同时实现 GO 的还原；此种技术路径的大规模制备需要对正极材料生产工艺和相关设备进行调整和优化。

6.1.4 氧化石墨烯作为导电添加剂在锂离子电池中应用

相对上述在原位制备正负极材料的过程中添加 GO 的方法，直接在制备电极时添加石墨烯作为导电剂材料更加简单易行。

在正常充放电情形下，正极材料需要脱嵌锂离子同时得失电子，为了发挥其性能，必须保证其离子和电子的传输通道。由于常用的几种正极材料几乎都是导电性较差的半导体，需要加入导电剂以增强其导电性能，从而提高电池的倍率和循环性能。一般认为，导电添加剂在锂离子电池电极中起两个作用：a.增强电子电导性；b.吸收和保持电解液溶液提高离子传导率。碳材料是锂离子电池中主要应用的导电剂，按导电剂材料形态可以分为颗粒状的炭黑导电剂、线状的碳纳米管以及片状的石墨烯导电剂等。由于过多添加不参与能量存储的导电剂会降低电极的能量密度，因此需要寻求添加量较少的高效导电添加剂。

石墨烯纳米片作为高导电率的二维柔性材料已被广泛研究证明可以提升正极材料的性能[14~16]。在利用石墨烯作为 $LiFePO_4$ 正极材料导电剂的研究中，研究人员提出石墨烯与正极材料颗粒的"点-面"接触导电模型，认为石墨烯是一种高效的锂离子电池导电剂[17]。相对于传统导电剂材料，石墨烯的高导电性和良好分散性，可以降低导电剂的使用量，从而提高活性物质的用量。而且石墨烯作为导电剂可以增强极片的吸液能力，减缓了活性材料的体积效应，使得电池的循环性能得到提高[18]。目前用于导电剂的石墨烯可以由两种方法制备得到：一种是将石墨直接进行物理法剥离得到石墨烯分散液；另一种是将 GO 先进行还原得到石墨烯，接着通过一定的分散工艺制备得到石墨烯分散液。李用等[19]人通过 GO 与炭黑超声混合后进行高温还原得到石墨烯和炭黑复合导电剂，添加在 $LiFePO_4$ 正极材料中，在 10C 倍率下的容量为 73mA·h/g，比单独使用商品化炭黑导电剂容量提高将近 25%。Tang 等[20]人将 GO 置于高真空低温环境中制备得到石墨烯，然后在 $LiCOO_2$ 中使用极少量的石墨烯和炭黑作为导电剂，建立了良好的导电网络，提高了电池的倍率性能和循环性能。在 5C 放电倍率下容量有 116.5mA·h/g，在 1C 倍率下循环 50 圈容量保持率达 96.4%。目前已有电池企业已经开始使用石墨烯作为 $LiFePO_4$ 正极材料中的导电剂，并且已在三元正极材料中试用。高坡等[21]人将石墨烯和碳管作为复合导电剂，形成"点-线-面"三维立体导电网络结构，使得电池内阻降低；在导电剂添加量为1%时，使用复合导电剂电池放电容量可比单独使用碳管提高7mA·h/g，在10C倍率放电下容量可达128mA·h/g。该研究中使用的石墨烯是将GO高温还原后通过一定的分散工艺得到的石墨烯分散液。

6.2 氧化石墨烯在超级电容器中的应用

电化学电容器通常被称为超级电容器，与传统的静电和电解电容器相比具有更高的电容值[22]。典型的超级电容器由两个电极组成，置于同一电解液中，两个电极之间被多孔膜隔离开。超级电容器于1978年首次在市场上推出，以提供计算机内存备份电源[23]。据报道空客A380紧急舱门上也应用了超级电容器[24]。超级电容器与锂电池在结构和制造上尽管有相似之处，但超级电容器特别是双电层电容器（Electrical Double-Layercapacitor，简称EDLCs）可以在100万个周期内以高电荷放电速率运行，并在几秒钟内储存和释放能量。因此，超级电容器具有大电流、高功率密度（10kW/kg）、长周期寿命（＞100倍电池寿命）和低维护成本等优势。另一方面，超级电容器的能量密度（4～5W·h/kg）远低于电池的能量密度（120～170W·h/kg）[25]。此外，超级电容器的电压随放电状态而变化，而电池的输出电压基本恒定[26]。因此超级电容器被认为是电池的补充，尤其是在高倍率的应用中，如电动汽车或混合动力电动汽车的能量回收装置中。

根据不同的储能机制[27]，超级电容器可分为双电层电容和赝电容。在双电层电容器中，电压施加在电极上通过静电吸附电解液离子，在电极上形成双电层电解液界面。目前，多孔碳质材料，包括活性炭、碳化物、模板碳、碳气凝胶、碳纤维、碳纳米管等，已被广泛应用于双电层电容器的电极材料[28]。高导电性、大比表面积的石墨烯也在双电层电容器领域受到广泛研究和应用。与双电层电容器相比，赝电容是通过电解液和电极之间可逆的氧化还原反应储存能量。导电聚合物和金属氧化物具有较高的电化学活性和较高的电子转移能力，是具有吸引力的赝电容材料。与双电层电容相比，赝电容可以表现出更高的电容和能量密度，但由于电极材料的低电导和电极表面反应中缓慢的电子转移动力学，功率密度相对较低。此外，在长期氧化还原反应过程中，电极的降解和老化可能会导致其寿命缩短，所以赝电容材料普遍存在稳定性差的问题。在这方面，石墨烯材料的引入可以增加导电性、提升比表面积、增加高机械强度和柔性，并可实现多孔微结构，因而提升赝电容的综合性能[29]。

6.2.1 氧化石墨烯在双电层电容器中的应用

有研究认为单层石墨烯的固有双层电容为$21mF/cm^2$，如果石墨烯的理论比表面积被完全利用[30]，则基于石墨烯的双电层电容器理论电容达550F/g。Vivekchand等[31]人较早报道了利用石墨烯作为双电层电容器电极的研究。其研究中将GO在1050℃热还原后作为电极材料，比表面积为925m^2/g。在1mol/LH$_2$SO$_4$电解液中，扫描速度为100mV/s时，测定的比电容为117F/g，高于单壁碳纳米管SWCNTs（64F/g）和多壁碳纳米管（MultiWalled Carbon Nanotubes，简称MWNTs）（14F/g）。此外，Stoller等[32]人通过水合肼还原GO得到的石墨烯粉体具有705m^2/g的比表面积，也被用作电极材料制备双电层电容器（Electrical Double-Layer Capacitor，简称EDLC）。在10mA/g的电流密度下，氢氧化钾水溶液和有机电解液中比电容分别为135F/g和99F/g；在40mV/s扫描速率下，比电容分别为107F/g和85F/g。Zhu等[33]人使用微波法将GO还原为石墨烯后作为电极材料。在KOH电解液中，150mA/g电流下比电容为191F/g，600mA/g电流下比电容仍可达174F/g。目前为止，从石墨烯双电层电容器中获得的实验结果通常小于200F/g，低于上述理论值。以下针对不同形貌的石墨烯材料作为双电层超级电容器的电极材料研究结果进行总结。

（1）石墨烯纸/薄膜电极

如前章节所述，石墨烯薄片之间具有的π-π相互作用会造成石墨烯微片的团聚或回叠，限制石墨烯电极的性能。此外，石墨烯的高比表面积会受到电解质浸润效果的影响，石墨烯的高导电性也可能受到层间接触电阻的影响。Wang等[34]人采用肼蒸汽还原氧化石墨烯的方法得到石墨烯纸，在KOH电解液中比电容高达205F/g，功率密度为10kW/kg，能量密度为28.5W·h/kg，并具有良好的循环性能，在1200周循环后电容可保留90%。Feng等[35]人提出了一种低温还原方法，使用钠-氨（Na-NH$_3$）体系在干燥的冰-丙酮浴中作为催化剂，将GO膜直接还原成石墨烯膜，比表面积为648m^2/g；作为双电层电容器电极材料时在10mV/s扫描速率下比电容为263F/g。Yang等[36]人采用准分子激光在不同的激光能量和辐照时间下还原GO。激光还原制备的石墨烯膜呈随机聚集、褶皱、无序、小片状的固体状态。激光处理后氧官能团总量明显减少，C-C/C-O强度比增加，C=O双键的含量增加。用激光还原石墨烯作为超级电容器的电极材料，在0.5M Na$_2$SO$_4$水溶液中、1.04A/g电流时比电容约为141F/g。Zhang等[37]人研究发现，KOH活化GO也可制备高导电性、高强度的多孔石墨烯薄膜，制备过程如图6.1所示。得到的材料具有很高的比表面积（2400m^2/g）和高的导电性（5880S/m）。使用这种电极材料的双电层电容器具有较低的等效串联电阻，在保持高比电容（120F/g）和能量密度（26W·h/kg）的同时，输出功率高达500kW/kg。

（2）三维石墨烯电极

三维互联的多孔结构，如水凝胶、气凝胶、泡沫、海绵等可以保持石墨烯薄片的固有特性[38]，还可以提供具有良好机械稳定性、高比表面积和高导电网络的电极，从而改善了存储电容、可循环性和倍率能力[39]。热还原方法被广泛用于生产3D多孔材料[40]。参照图6.2，通过水热法可以低成本、短时间得到所需的宏观石

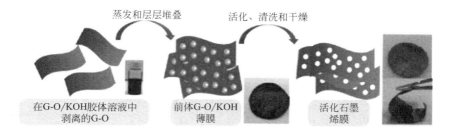

图 6.1 活化石墨烯膜的制备过程[37]

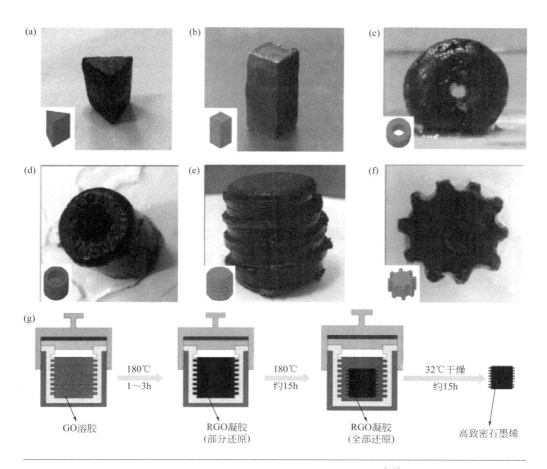

图 6.2 不同模型石墨烯水凝胶及其制备示意[41]

墨烯材料[41]。例如,Kou 等[42]人用一种快速的方法制备了多孔石墨烯电极,有效提高了储能性能。研究人员采用一种基于 GO 液晶行为的湿法自旋组装方法,可直接作为无黏结性超级电容器电极使用。具有"开孔"的石墨烯水凝胶薄膜连续自旋，所得到的超级电容器电极比无序石墨烯片表现出更好的电化学性能。该石墨烯水凝胶膜在 1A/g 下比电容最高达 203F/g，在 50A/g 下仍保持了 67.1％的电容（140F/g）。据报道，Zhou 等[43]人利用还原氧化石墨烯得到了石墨烯凝胶，作为超级电容器电极使用时，在 1A/g 下测得 140F/g 的比电容，具有较高能量密度（15.4W·h/kg）和功率密度 16.3kW/kg（30A/g）。

（3）碳质混合电极

将碳纳米管（Carbon nanotube，简称 CNTs）和石墨烯复合可以形成 3D 网络[44]。研究发现，由还原的 GO 和 CNTs（重量为 9∶1）组成的混合电极可以体现较高的比表面积（538.9m^2/g）和比电容（20mV/s 下为 326.5F/g），能量和功率密度分别高达 21.74W·h/kg 和 78.29kW/kg。Ramesh 等[45]人制备了 GO/氮掺杂碳纳米管/纤维素复合材料作为超级电容器电极，目的是在石墨烯层上复合碳纳米管，以避免石墨烯层的自聚集，同时实现高能量密度、高电容和功率密度。得到的 RGO/碳纳米管/纤维素复合材料在 6A/g 电流下的电容为 264F/g，具有 36.66kW/kg 的功率密度。碳球也可以在石墨烯薄片之间插入，以形成具有更好电容性能的三维结构[46]。这些碳基复合材料为双电层电容提供了优异性能的电极材料，例如高比表面积、高导电以及快速的电子和电解质离子传输等特性[47]。由于不同碳纳米材料中的本征电容值有所不同[48]，所以混合材料的比电容会根据不同的复合情况有所不同，仍需深入研究。此外，虽然由于碳材料的密度较低，因此具有较好的质量比电容性能，但如果电极密度过低将会大大损失体积比性能，这是碳基材料超级电容器研究中值得注意的问题。为进一步提升体积比性能，需要发展更先进的材料制备技术，并研究密度/孔隙率/比表面积和电容性能之间的关系[49]。

6.2.2 氧化石墨烯在赝电容器中的应用

（1）导电聚合物/石墨烯复合材料

在导电聚合物中，聚苯胺（Polyaniline，简称 PANI）因其理论电容高（2000F/g）、低成本和易于合成而被广泛应用于超级电容器[50]。石墨烯/PANI 复合材料也已由多种方法，如原位聚合法[51]、物理混合法[52]、层层自组装法[53]、CVD 法[54]、电纺法[55]和化学接枝法[56]等制备出来。Xu 等[57]人报道了一种由原位化学聚合制备的 GO/PANI 复合材料，其中 PANI 纳米线垂直生长在 GO 表面形成一种分层结构的纳米复合材料。该复合材料在 0.2A/g 时表现出较高的比电容（555F/g），在 2A/g 的大电流密度下比电容为 227F/g；在 2000 次循环后，电容器还有 92％的电容保持率，而纯 PANI 则只保存有 74％的初始电容值。Cong 等[58]人开发了一种方法，可以通过一步电化学沉积法在石墨烯纸表面生长 PANI 纳米棒从而得到柔性石墨烯/PANI 复合薄膜材料。该复合材料在 1A/g 电流密度下比电容高达763F/g，而单独的石墨烯纸和PANI膜比电容分别仅有180F/g和520F/g。该复合材料作为电极材料还具有良好的循环性能，在循环1000圈后其比电容还可保留原始比电容的82%。

石墨烯/聚吡咯（Polypyrrole，简称PPY）复合材料也可通过电纺聚合[59]、化学聚合[60]、电化学沉积[61]和物理混合[62]等方法得到。与单独的PPY相比，所有的复合材料作为电极材料的电极都提升了超级电容器性能。其他导电聚合物，如聚乙烯二氧噻吩［Poly（3,4-Ethylenedioxythiophene），简称PEDOT］，虽然成本较高，也研究和报道。在最近的一项研究中，Zhang等[63]人先将GO进行还原，接着通过原位聚合将PANI、PPY和PEDOT分别聚合在石墨烯表面得到石墨烯与各种导电聚合物的复合材料。石墨烯/PANI、石墨烯/PPY和石墨烯/PEDOT复合材料在0.3A/g电流密度下的比电容分别为361F/g、248F/g和108F/g。与其他两种复合材料相比，石墨烯/PANI复合材料具有更高的比电容，在循环1000圈后，比电容仍可保留原始电容的80%以上。这可能是由于二者之间的协同作用，石墨烯提高了复合材料的导电性，并在长循环过程中抑制了其体积膨胀和收缩，同时导电聚合物也抑制了石墨烯材料的自团聚。

（2）金属氧化物/石墨烯复合材料

由于成本低和电容值高[64]，金属氧化物长期以来一直被作为赝电容器的潜在电极材料研究。但是，金属氧化物的导电性较低，基于金属氧化物材料制备的赝电容具有较差的循环和倍率性能。石墨烯可与金属氧化物复合，改善其导电性和机械/电化学稳定性。

在金属氧化物赝电容电极材料中，RuO_2由于其具有高理论电容值（1358F/g）和相对较高导电性（300S/cm）而成为一种有前途的电极材料[65]。RuO_2作为超级电容电极材料在酸性电解液中也具有较宽的电压窗口，从而使其具有较高的能量密度[66]。然而，RuO_2纳米颗粒容易形成大量的聚集物，导致氧化—还原反应不完全，电化学循环性能显著下降。Hu等[67]人先将GO与预处理的CNT进行化学还原复合，接着通过阳极沉积方法制备三维石墨烯/CNT/RuO_2纳米复合材料。该复合材料在未还原时作为电极材料其比电容可达808F/g，经过150℃还原后其在25mV/s下比电容高达973F/g，远高于RuO_2纳米颗粒的比电容（435F/g）；当扫描速度从5mV/s增加到500mV/s时容量保持率为60.5%。虽然RuO_2对于实际应用成本过高且具有毒性，但长期以来一直被用作探索超电容性能的一种模型材料。

二氧化锰通常被认为是超级电容器的一种可实用化的电极材料，因为它具有高理论电容（1380F/g）、低成本、储量大、环境友好和易于大规模生产等优势[68]。但是，它的低电导率（$10^{-5} \sim 10^{-6}$S/cm）限制了其高功率性能，影响了它的进一步应用。Naderi等[69]人通过水热法制备得到氮掺杂石墨烯，接着采用超声法制备得到氮掺杂石墨烯/二氧化锰复合材料，该复合材料中二氧化锰颗粒均匀分布在氮掺杂石墨烯表面。其在2mV/s扫速下比电容为522F/g，于200mV/s扫速下循环

4000圈后比电容仍可保持原始容量的96.3%,作者认为该复合材料高比电容和优良的循环稳定性是由于二者之间的协同作用。

除了RuO_2[70]和MnO_2[71]之外,有诸多研究将石墨烯与其他金属化合物如ZnO[72]、SnO_2[73]、CeO_2[74]、Mn_3O_4[75]、Fe_3O_4[76]、Co_3O_4[77]、V_2O_5[78]、$Co(OH)_2$[79]和$Ni(OH)_2$[80]等进行复合制备电极材料。在这些复合材料中,金属氧化物通过丰富的表面/近表面氧化还原反应提供赝电容,而石墨烯则通过表面离子的吸附/解吸提供双电层电容。与此同时,石墨烯还作为支撑金属氧化物颗粒的坚固框架并形成一种导电网络,提高电解质离子和电子的运输速率。

6.3 氧化石墨烯在燃料电池中的应用

燃料电池是一种电化学装置,通过氧化—还原反应将化学能转化为电能,具有能量密度高、效率高和污染物排放低等优点。基于电池中使用的电解质不同,燃料电池通常包括五种类型,分别是:碱性燃料电池(Alkaline Fuel Cell,简称AFC)、质子交换膜燃料电池(Proton Exchange Membrane Fuel Cell,简称PEMFC)、磷酸燃料电池(Phosphoric Acid Fuel Cell,简称PAFC)、隔膜和熔融碳酸燃料电池(Molten Carbonate Fuel Cell,简称MCFC)和固体氧化物燃料电池(Solid Oxide Fuel Cell,简称SOFC)[81]。

在燃料电池中,化学反应通常是阳极上燃料(如氢、甲醇、甲酸)被催化氧化,电子通过导线形成电流,离子通过电解液到达阴极。Pt一直被认为是阳极和阴极反应的最佳催化剂,尤其是低温燃料电池。其他稀有金属如Au、Ru、Pd、Ni和它们的合金也被报道用于催化燃料电池中的电极反应[82]。但贵金属催化剂存在反应动力学缓慢、成本高和交叉效应等问题[83]。而且催化剂通常要求在载体上均匀分散,以增加电活性表面积、提高循环稳定性和寿命。石墨烯大的比表面积可以分散催化剂纳米粒子,从而抑制粒子间的聚集,使更多有效的电活性位点暴露出来。特别是GO不仅含有孔洞、边缘褶皱等缺陷,还含有大量的官能团可以成为催化剂纳米粒子生长和铆定的基底[84]。Xin等[85]人通过一步还原法将H_2PtCl_6和GO还原制备得到铂/石墨烯复合催化剂(Pt/G),其中铂纳米粒子均匀地沉积在石墨烯薄片上。该复合催化剂对甲醇氧化反应和氧气还原反应显示出较高的催化活性;经300℃下还原2h后该复合催化剂催化能力相比铂/炭黑催化剂提高3.5倍。Iwan等[86]人先通过改进的Hummers'方法得到GO,并将其作为聚合物燃料电池中的阳极添加剂。添加GO的阳极其催化剂可以减少30%,在$374mA/cm^2$电流密度下,功率密度达$134mW/cm^2$。最近,石墨烯负载氧化物如钴氧化物[87]、锰氧化物[88]、铁氧化物、铜氧化物[89]和多金属氧化物[90]也被证明是有效的催化剂,具有良好的活性和催化稳定性。

此外，在燃料电池运行过程中，质子和水的输送要通过电解质膜进行。聚合物电解质膜（Polymer Electrolyte Membranes，简称 PEMs）要求具有对催化剂和各种中间体的耐受性、高的热传导和机械稳定性等[91]。目前的 PEMs 主要是基于聚四氟乙烯聚合物，但是成本高、电导率低使其应用受到限制。GO 由于具有大比表面积、高机械强度、良好的化学稳定性等优势，可用于开发高性能的 PEMs 以提高其机械和热稳定性、降低甲醇交换势垒和提升质子电导率等。Hong 等[92] 人将 0.5%（质量）磺化石墨烯在 DMF 中与聚四氟乙烯混合制备得到磺化石墨烯/聚四氟乙烯复合膜。相比于单独的聚四氟乙烯膜和 GO/SGON 复合膜，该磺化石墨烯/聚四氟乙烯复合膜在较宽温度范围内可以显著提高质子电导率并减少甲醇渗透率。以 1M 甲醇为阳极、氧气为阴极，在 60℃ 的测试温度下，该磺化石墨烯/聚四氟乙烯复合膜电池功率密度为 $132mW/cm^2$，而单独的聚四氟乙烯膜和 GO/SGON 复合膜电池功率密度分别为 $101mW/cm^2$ 和 $120mW/cm^2$。

6.4 氧化石墨烯在太阳能电池中的应用

太阳能电池或光伏电池可以将太阳能直接转化为电能。典型的光伏装置通常由阳极、阴极和之间的活性材料层组成，其中阴极是透明的，以便阳光能够通过。目前，太阳能电池商业应用的关键在于提高光电转换效率，同时通过开发高性能的活性层和电极材料来降低成本。作为碳原子以 sp^2 杂化形成的二维晶体[93]，石墨烯室温下电子迁移率约为 $20000cm^2/(V·s)$[94]，单层吸光率只有 2.3%[95]，使石墨烯及其衍生材料被广泛应用于透明电极[96]、对电极[97] 和电荷传输层等结构。此外，相比于 ITO 透明导电电极，石墨烯具有更好的机械强度及韧性，使其在柔性光伏器件的透明电极中具有更好潜力[98]。

Xu 等[99] 人将 GO 溶液旋涂成膜，然后在 700℃ 下用肼蒸汽还原，所得石墨烯薄膜的薄层电阻为 $1.79×10^4 \Omega/\square$，电导率为 $22.3S/cm$，将其在有机光伏电池中（Organic Photovoltaic Cell，简称 OPV）作为透明电极，所得器件的光电转换效率为 0.13%。这种方法制备得到的石墨烯薄膜不仅可以用于有机光伏电池，还可以用于其他光学器件，例如平板显示器等。Zhang 等[100] 人对 GO 在 950℃ 下进行热还原，再使用标准工业光刻以及 O_2 等离子体蚀刻工艺对还原的石墨烯薄膜进行可控刻蚀，制备了石墨烯网状透明电极，提高其透光率到 65%，同时面电阻降至 $750\Omega/\square$（未经刻蚀的还原石墨烯薄膜为 $17.9k\Omega/\square$），将其用于有机光伏电池中光电转换效率可达到 2.04%，但仍低于 ITO 作电极的光伏器件功率转换效率（3%），可能是透光率仍然较低，以及电极内部结构缺陷导致材料导电性较差所致。

石墨烯还可以用于其他光电器件中，例如染料敏化太阳能电池（Dye Sensitized Solar Cell，简称 DSSC）。由 Gratzel[101] 首创的 DSSC 被认为是传统硅基电池的一种极具吸引力的替代品，具有高光电转换效率（>10%）、低生产成本和工艺相对简单的优势[102]。典型的 DSSC 设备主要由纳米多孔半导体薄膜、染料敏化剂、氧化还原电解质、对电极和导电基底等几部分组成[103]。2008 年，Wang 等[104] 人提出将 GO 进行热还原制备得到的石墨烯作为光阳极应用在 DSSC 中。该电极的导电率为 550S/cm，还具有高透明、表面光滑、热稳定性好等优点，但是该电极的 DSSC 光电转换效率仅有 0.26%。Kim 等[105] 人利用紫外光还原氧化石墨烯的方法制备了石墨烯和 TiO_2 复合物，并将该复合物作为 DSSC 中的光阳极，石墨烯的存在使电池的功率转换效率从 4.89% 增加到 5.26%。作者认为，功率转换效率的提高是由于石墨烯层的存在阻止了 I_3^- 离子与氟掺杂的锡氧化物（FTO）导电玻璃的直接接触，降低了电子从 FTO 回传到 I_3^- 离子的概率。Sun 等[106] 人先将 GO 通过联胺还原，制备得到石墨烯/TiO_2 复合物，将其作为光阳极应用在 DSSC 中，功率转换效率为 4.28%，比单独 TiO_2 光阳极提高 59%。

石墨烯还可用于 DSSC 中对电极部分。对电极是 DSSC 的重要组成部分，催化着电解液中 I_3^- 的还原反应。Pt 由于其优异的催化性能和导电性能在 DSSC 中作为广泛使用的催化剂材料，但价格比较昂贵。Hong 等[107] 人先使用改进的 Hummers' 法制备 GO 分散液，接着制备功能化石墨烯/聚（3,4-乙烯二氧噻吩）-聚苯乙烯磺酸（PEDOT-PSS）复合材料，并将其通过旋涂工艺沉积于 ITO 导电玻璃上而作为 DSSC 中的对电极。涂层（含有 1% 石墨烯）膜厚为 60nm，该复合物作为对电极具有较高的透光度（>80%）和催化活性，功率转化效率达到 4.5%。使用 Pt/石墨烯复合材料作为 DSSC 对电极的优势在于保持了功率转换效率的同时降低了 Pt 的含量，从而降低了 DSSC 的成本，有利于实现其产业化应用。但目前 Pt/石墨烯复合对电极的制备工艺仍然较为复杂、成本较高，更为简单和稳定的合成方法将有可能成为 Pt/石墨烯复合对电极产业化的关键技术。

6.5　小结

GO 及其衍生物、复合物在能源领域得到了较大关注。本章主要介绍了以 GO 为前驱体制备得到石墨烯及其复合材料在锂离子电池电极材料、超级电容器材料、燃料电池材料和太阳能电池材料中的研究进展。目前石墨烯在锂离子电池中作为正极材料的导电剂已经实现商业化，石墨烯在其他能源领域（超级电容器、燃料电池、太阳能电池）也有很大的研究进展。相信通过进一步研究，石墨烯可以更好地促进新能源领域的快速提升，对提高能源的利用效率、节能减排及环境保护有重要的作用。

参考文献

[1] Novoselov K. S., Geim A. K., Morozov S, et al. Electric field effect in atomically thin carbon films [J]. Science, **2004**, 306 (5696): 666-669.

[2] Bulusheva Lyubov, Stolyarova Svetlana, Chuvilin Andrey, et al. Creation of nanosized holes in graphene planes for improvement of rate capability of lithium-ion batteries [J]. Nanotechnology, **2018**, 29 (13): 134001.

[3] Lian Peichao, Zhu Xuefeng, Liang Shuzhao, et al. Large reversible capacity of high quality graphene sheets as an anode material for lithium-ion batteries [J]. Electrochimica Acta, **2010**, 55 (12): 3909-3914.

[4] He Yu Shi, Gao Pengfei, Chen Jun, et al. A novel bath lily-like graphene sheet-wrapped nano-Si composite as a high performance anode material for Li-ion batteries [J]. Rsc Advances, **2011**, 1 (6): 958-960.

[5] Lee Jeong K, Smith Kurt B, Hayner Cary M, et al. Silicon nanoparticles-graphene paper composites for Li ion battery anodes [J]. Chemical Communications, **2010**, 46 (12): 2025-2027.

[6] Li Qiuli, Chen Dingqiong, Li Kun, et al. Electrostatic self-assembly bmSi@C/rGO composite as anode material for lithium ion battery [J]. Electrochimica Acta, **2016**, 202: 140-146.

[7] Deosarkar Manik P., Pawar M. S., Bhanvase Bharat A. In situ sonochemical synthesis of Fe_3O_4-graphene nanocomposite for lithium rechargeable batteries [J]. Chemical Engineering Processing Process Intensification, **2014**, 83 (9): 49-55.

[8] Wu Zhong-Shuai, Ren Wencai, Wen Lei, et al. Graphene anchored with co (3) o (4) nanoparticles as anode of lithium ion batteries with enhanced reversible capacity and cyclic performance [J]. Acs Nano, **2010**, 4 (6): 3187-3194.

[9] Hao Shiji, Zhang Bowei, Wang Ying, et al. Hierarchical three-dimensional Fe_3O_4@porous carbon matrix/graphene anodes for high performance lithium ion batteries [J]. Electrochimica Acta, **2017**, 260: 965-973.

[10] Wang Li, Wang Haibo, Liu Zhihong, et al. A facile method of preparing mixed conducting $LiFePO_4$/graphene composites for lithium-ion batteries [J]. Solid State Ionics, **2010**, 181 (37): 1685-1689.

[11] Ding Y. H., Jiang Yolanda, Xu Fu, et al. Preparation of nano-structured LiFePO/graphene composites by co-precipitation method [J]. Electrochemistry Communications, **2010**, 12 (1): 10-13.

[12] Wang Xiuling, Dong Shuling, Wang Hongxia. Design and fabrication of N-doped graphene decorated $LiFePO_4$@C composite as a potential cathode for electrochemical energy storage [J]. Ceramics International, **2018**, 44 (1): 464-470.

[13] Rao Chitturi Venkateswara, Reddy Arava Leela Mohana, Ishikawa Yasuyuki, et al. $LiNi_{1/3}Co_{1/3}Mn_{1/3}O_2$-Graphene Composite as a Promising Cathode for Lithium-Ion Batteries [J]. Acs Applied Materials Interfaces, **2011**, 3 (8): 2966-2972.

[14] Su Fang Yuan, He Yan Bing, Li Baohua, et al. Could graphene construct an effective conducting network in a high-power lithium ion battery? [J]. Nano Energy, **2012**, 1 (3): 429-439.

[15] Ha Jeonghyun, Park Seung-Keun, Yu Seung-Ho, et al. A chemically activated graphene-encapsulated $LiFePO_4$ composite for high-performance lithium ion batteries [J]. Nanoscale, **2013**, 5 (18): 8647-8655.

[16] Zhang Biao, Yu Yang, Liu Yusi, et al. Percolation threshold of graphene nanosheets as conductive additives in $Li_4Ti_5O_{12}$ anodes of Li-ion batteries [J]. Nanoscale, **2013**, 5 (5): 2100-2106.

[17] Su Fang-Yuan, You Conghui, He Yan-Bing, et al. Flexible and planar graphene conductive additives for lithium-ion batteries [J]. Journal of Materials Chemistry, **2010**, 20 (43): 9644-9650.

[18] Ke Lei, Lv Wei, Su Fang Yuan, et al. Electrode thickness control: Precondition for quite different functions of graphene conductive additives in LiFePO$_4$ electrode [J]. Carbon, **2015**, 92: 311-317.

[19] 李用, 吕小慧, 苏方远, 等. 石墨烯/炭黑杂化材料: 新型、高效锂离子电池二元导电剂 [J]. 新型炭材料, **2015**, 30 (2): 128-132.

[20] Tang Rui, Yun Qinbai, Lv Wei, et al. How a very trace amount of graphene additive works for constructing an efficient conductive network in LiCoO$_2$-based lithium-ion batteries [J]. Carbon, **2016**, 103: 356-362.

[21] 高坡, 张彦林, 颜健. 石墨烯/碳纳米管复合导电剂对 LiNi$_{1/3}$Co$_{1/3}$Mn$_{1/3}$O$_2$ 的影响 [J]. 电池, **2017**, 6: 008.

[22] Martin Winter, Brodd Ralph J. What are batteries, fuel cells, and supercapacitors? [J]. Cheminform, **2004**, 35 (50): 4245.

[23] Miller John R., Simon Petrice. Electrochemical Capacitors for Energy Management [J]. Science, **2008**, 321 (5889): 651-652.

[24] Simon Patrice, Gogotsi Yury. Materials for electrochemical capacitors [J]. Nature Materials, **2008**, 7 (11): 845-854.

[25] Burke Andrew. R&D considerations for the performance and application of electrochemical capacitors [J]. Electrochimica Acta, **2008**, 53 (3): 1083-1091.

[26] Whittingham M Stanley. Materials challenges facing electrical energy storage [J]. Mrs Bulletin, **2008**, 33 (4): 411-419.

[27] Choi Hyun Jung, Jung Sun Min, Seo Jeong Min, et al. Graphene for energy conversion and storage in fuel cells and supercapacitors [J]. Nano Energy, **2012**, 1 (4): 534-551.

[28] Pandolfo A. G., Hollenkamp A. F. Carbon properties and their role in supercapacitors☆ [J]. Journal of Power Sources, **2006**, 157 (1): 11-27.

[29] Zhang Li Li, Zhou Rui, Zhao XS. Graphene-based materials as supercapacitor electrodes [J]. Journal of Materials Chemistry, **2010**, 20 (29): 5983-5992.

[30] Xia Jilin, Chen Fang, Li Jinghong, et al. Measurement of the quantum capacitance of graphene [J]. Nature Nanotechnology, **2009**, 4 (8): 505-509.

[31] Vivekchand S. R. C., Rout Chandra Sekhar, Subrahmanyam K. S., et al. Graphene-based electrochemical supercapacitors [J]. Journal of Chemical Sciences, **2008**, 120 (1): 9-13.

[32] Stoller Meryl D., Park Sungjin, Zhu Yanwu, et al. Graphene-based ultracapacitors [J]. Nano Letters, **2008**, 8 (10): 3498-3502.

[33] Zhu Yanwu, Murali Shanthi, Stoller Meryl D., et al. Microwave assisted exfoliation and reduction of graphite oxide for ultracapacitors [J]. Carbon, **2010**, 48 (7): 2118-2122.

[34] Wang Yan, Shi Zhiqiang, Huang Yi, et al. Supercapacitor Devices Based on Graphene Materials [J]. Journal of Physical Chemistry C, **2009**, 113 (30): 13103-13107.

[35] Feng Hongbin, Cheng Rui, Zhao Xin, et al. Corrigendum: A low-temperature method to produce highly reduced graphene oxide [J]. Nature Communications, **2013**, 4 (2): 1539.

[36] Yang Dongfang, Bock Christina. Laser reduced graphene for supercapacitor applications [J]. Journal of Power Sources, **2017**, 337: 73-81.

[37] Zhang Lili, Zhao Xin, D Stoller Meryl, et al. Highly conductive and porous activated reduced graphene oxide films for high-power supercapacitors [J]. Nano Letters, **2012**, 12 (4): 1806-1812.

[38] Stefania Nardecchia, Daniel Carriazo, M Luisa Ferrer, et al. Three dimensional macroporous architectures and aerogels built of carbon nanotubes and/or graphene: synthesis and applications [J]. Chemical Society Reviews, **2013**, 42 (2): 794-830.

[39] Tiwari Jitendra N, Tiwari Rajanish N, Kim Kwang S. Zero-dimensional, one-dimensional, two-dimensional and three-dimensional nanostructured materials for advanced electrochemical energy devices [J]. Progress in Materials Science, **2012**, 57 (4): 724-803.

[40] Li Chun, Shi Gaoquan. Three-dimensional graphene architectures [J]. Nanoscale, **2012**, 4 (18): 5549-5563.

[41] Bi Hengchang, Yin Kuibo, Xie Xiao, et al. Low temperature casting of graphene with high compressive strength [J]. Advanced Materials, **2012**, 24 (37): 5123-5123.

[42] Kou Liang, Liu Zheng, Huang Tieqi, et al. Wet-spun, porous, orientational graphene hydrogel films for high-performance supercapacitor electrodes [J]. Nanoscale, **2015**, 7 (9): 4080-4087.

[43] Zhou Qinqin, Gao Jian, Li Chun, et al. Composite organogels of graphene and activated carbon for electrochemical capacitors [J]. Journal of Materials Chemistry A, **2013**, 1 (32): 9196-9201.

[44] Yang Shin Yi, Chang Kuo Hsin, Tien Hsi Wen, et al. Design and tailoring of a hierarchical graphene-carbon nanotube architecture for supercapacitors [J]. Journal of Materials Chemistry, **2011**, 21 (7): 2374-2380.

[45] Ramesh Sivalingam, Sivasamy Arumugam, Kim Heung Soo, et al. High-performance N-doped MWCNT/GO/cellulose hybrid composites for supercapacitor electrodes [J]. Rsc Advances, **2017**, 7 (78): 49799-49809.

[46] Guo Chun Xian, Li Chang Ming. A self-assembled hierarchical nanostructure comprising carbon spheres and graphene nanosheets for enhanced supercapacitor performance [J]. Energy Environmental Science, **2011**, 4 (11): 4504-4507.

[47] Lin Lu-Yin, Yeh Min-Hsin, Tsai Jin-Ting, et al. A novel core-shell multi-walled carbon nanotube@ graphene oxide nanoribbon heterostructure as a potential supercapacitor material [J]. Journal of Materials Chemistry A, **2013**, 1 (37): 11237-11245.

[48] Liu Ruili, Wan Li, Liu Shaoqing, et al. Supercapacitors: An Interface-Induced Co-Assembly Approach Towards Ordered Mesoporous Carbon/Graphene Aerogel for High-Performance Supercapacitors (Adv. Funct. Mater. 4/2015) [J]. Advanced Functional Materials, **2015**, 25 (4): 526-533.

[49] 叶江林, 朱彦武. 氢氧化钾活化制备超级电容器多孔碳电极材料 [J]. 电化学, **2017**, (5): 548-559.

[50] Nanjundan Ashok Kumar, Jong-Beom Baek. Electrochemical supercapacitors from conducting polyaniline-graphene platforms [J]. Chemical Communications, **2014**, 50 (48): 6298-6308.

[51] Sakthivel S, Boopathi A. Characterization Study of Conducting Polymer PANI-GO Nano Composite Thin Film Supercapacitor [J]. Journal of Pure Applied Industrial Physics, **2016**, 6 (3): 34-41.

[52] Li Zhefei, Zhang Hangyu, Liu Qi, et al. Fabrication of high-surface-area graphene/polyaniline nanocomposites and their application in supercapacitors [J]. Acs Applied Materials Interfaces, **2013**, 5 (7): 2685-2691.

[53] Lu Jinlin, Liu Wanshuang, Ling Han, et al. Layer-by-layer assembled sulfonated-graphene/polyaniline nanocomposite films: enhanced electrical and ionic conductivities, and electrochromic properties [J]. Rsc Advances, **2012**, 2 (28): 10537-10543.

[54] Xue Mianqi, Li Fengwang, Zhu Juan, et al. Structure-Based Enhanced Capacitance: In Situ Growth of Highly Ordered Polyaniline Nanorods on Reduced Graphene Oxide Patterns [J]. Advanced Functional Materials, **2012**, 22 (6): 1284-1290.

[55] Huang Tieqi, Zheng Bingna, Kou Liang, et al. Flexible high performance wet-spun graphene fiber supercapacitors [J]. Rsc Advances, **2013**, 3 (46): 23957-23962.

[56] Nanjundan Ashok Kumar, Hyun-Jung Choi, Ran Shin Yeon, et al. Polyaniline-grafted reduced graphene oxide for efficient electrochemical supercapacitors [J]. Acs Nano, **2012**, 6 (2): 1715-1723.

[57] Xu Jingjing, Wang Kai, Zu Sheng-Zhen, et al. Hierarchical nanocomposites of polyaniline nanowire arrays on graphene oxide sheets with synergistic effect for energy storage [J]. Acs Nano, **2010**, 4 (9): 5019-5026.

[58] Cong Huai-Ping, Ren Xiao-Chen, Wang Ping, et al. Flexible graphene-polyaniline composite paper for high-performance supercapacitor [J]. Energy Environmental Science, **2013**, 6 (4): 1185-1191.

[59] Österholm Anna, Lindfors Tom, Kauppila Jussi, et al. Electrochemical incorporation of graphene oxide into conducting polymer films [J]. Electrochimica Acta, **2012**, 83 (12): 463-470.

[60] Wu Wenling, Yang Liuqing, Chen Suli, et al. Core-shell nanospherical polypyrrole/graphene oxide composites for high performance supercapacitors [J]. RSC Advances, **2015**, 5 (111): 91645-91653.

[61] Zhou Haihan, Han Gaoyi, Xiao Yaoming, et al. Facile preparation of polypyrrole/graphene oxide nanocomposites with large areal capacitance using electrochemical codeposition forsupercapacitors [J]. Journal of Power Sources, **2014**, 263 (263): 259-267.

[62] Zhang Jing, Yu Yao, Liu Lin, et al. Graphene-hollow PPy sphere 3D-nanoarchitecture with enhanced electrochemical performance [J]. Nanoscale, **2013**, 5 (7): 3052-3057.

[63] Zhang Jintao, Zhao X. S. Conducting Polymers Directly Coated on Reduced Graphene Oxide Sheets as High-Performance Supercapacitor Electrodes [J]. Journal of Physical Chemistry C, **2012**, 116 (9): 5420-5426.

[64] Jiang Jian, Li Yuanyuan, Liu Jinping, et al. Recent advances in metal oxide-based electrode architecture design for electrochemical energy storage [J]. Advanced Materials, **2012**, 24 (38): 5166-5180.

[65] Ozolins Vidvuds, Zhou Fei, Asta Mark. Ruthenia-based electrochemical supercapacitors: insights from first-principles calculations [J]. Accounts of Chemical Research, **2013**, 46 (5): 1084-1093.

[66] Chen Luyang, Hou Y, Kang Jianli, et al. Toward the Theoretical Capacitance of RuO_2 Reinforced by Highly Conductive Nanoporous Gold [J]. Advanced Energy Materials, **2013**, 3 (7): 851-856.

[67] Hu Chi Chang, Wang Chia Wei, Chang Kuo Hsin, et al. Anodic composite deposition of RuO_2/reduced graphene oxide/carbon nanotube for advanced supercapacitors [J]. Nanotechnology, **2015**, 26 (27): 274004.

[68] Bélanger Daniel, Brousse L, Long Jeffrey W. Manganese oxides: battery materials make the leap to electrochemical capacitors [J]. The Electrochemical Society Interface, **2008**, 17 (1): 49-52.

[69] Naderi Hamid Reza, Norouzi Parviz, Ganjali Mohammad Reza. Electrochemical Study of a Novel High Performance Supercapacitor Based on MnO_2/Nitrogen-Doped Graphene Nanocomposite [J]. Applied Surface Science, **2016**, 366: 552-560.

[70] Choi Bong, Huh Yun Suk, Hong Won, et al. Electroactive nanoparticle directed assembly of functionalized graphene nanosheets into hierarchical structures with hybrid compositions for flexible supercapacitors [J]. Nanoscale, **2013**, 5 (9): 3976-3981.

[71] Liu Mingkai, Tjiu Weng Weei, Pan Jisheng, et al. One-step synthesis of graphene nanoribbon-MnO_2 hybrids and their all-solid-state asymmetric supercapacitors [J]. Nanoscale, **2014**, 6 (8): 4233-4242.

[72] Saranya Murugan, Ramachandran Rajendran, Wang Fei. Graphene- Zinc oxide (G-ZnO) nanocomposite for Electrochemical Supercapacitor Applications [J]. Journal of Science Advanced Materials Devices, **2016**, 1 (4): 454-460.

[73] Chen Mingxi, Wang Huan, Li Lingzhi, et al. Novel and facile method, dynamic self-assemble, to prepare SnO$_2$/rGO droplet aerogel with complex morphologies and their application in supercapacitors [J]. Acs Applied Materials Interfaces, **2014**, 6 (16): 14327-14337.

[74] Dezfuli Amin Shiralizadeh, Ganjali Mohammad Reza, Naderi Hamid Reza. Anchoring Samarium Oxide Nanoparticles on Reduced Graphene Oxide for High-Performance Supercapacitor [J]. Applied Surface Science, **2017**, 402: 245-253.

[75] Qu Jiangying, Gao Feng, Zhou Quan, et al. Highly atom-economic synthesis of graphene/Mn$_3$O$_4$ hybrid composites for electrochemical supercapacitors [J]. Nanoscale, **2013**, 5 (7): 2999-3005.

[76] Shi Wenhui, Zhu Jixin, Sim Dao Haó, et al. Achieving high specific charge capacitances in Fe$_3$O$_4$/reduced graphene oxide nanocomposites [J]. Journal of Materials Chemistry, **2011**, 21 (10): 3422-3427.

[77] Xiang Chengcheng, Li Ming, Zhi Mingjia, et al. A reduced graphene oxide/Co$_3$O$_4$ composite for supercapacitor electrode [J]. Journal of Power Sources, **2013**, 226 (6): 65-70.

[78] Wu Yingjie, Gao Guohua, Wu Guangming. Self-assembled three-dimensional hierarchical porous V$_2$O$_5$/graphene hybrid aerogels for supercapacitors with high energy density and long cycle life [J]. Journal of Materials Chemistry A, **2015**, 3 (5): 1828-1832.

[79] Gao Shan, Sun Yongfu, Lei Fengcai, et al. Ultrahigh energy density realized by a single-layer β-Co (OH)$_2$ all-solid-state asymmetric supercapacitor [J]. Angewandte Chemie, **2014**, 53 (47): 12789-12793.

[80] Raj C. Retna, Bag Sourav. Layered Inorganic-Organic Hybrid Material Based on Reduced Graphene Oxide and α-Ni (OH)$_2$ for High Performance Supercapacitor Electrodes [J]. Journal of Materials Chemistry A, **2014**, 2 (42): 17848-17856.

[81] Steele Brian CH, Heinzel Angelika. Materials for fuel-cell technologies, in Materials For Sustainable Energy: A Collection of Peer-Reviewed Research and Review Articles from Nature Publishing Group [C]. World Scientific, **2011**, 224-231.

[82] Guo Shaojun, Wang Erkang. Noble metal nanomaterials: Controllable synthesis and application in fuel cells and analytical sensors [J]. Nano Today, **2011**, 6 (3): 240-264.

[83] Stamenkovic Vojislav R, Fowler Ben, Mun Bongjin Simon, et al. Improved oxygen reduction activity on Pt$_3$Ni (111) via increased surface site availability [J]. science, **2007**, 315 (5811): 493-497.

[84] Yang Ying-Kui, He Cheng-En, He Wen-Jie, et al. Reduction of silver nanoparticles onto graphene oxide nanosheets with N, N-dimethylformamide and SERS activities of GO/Ag composites [J]. Journal of Nanoparticle Research, **2011**, 13 (10): 5571.

[85] Xin Yuchen, Liu Jian-guo, Zhou Yong, et al. Preparation and characterization of Pt supported on graphene with enhanced electrocatalytic activity in fuel cell [J]. Journal of Power Sources, **2011**, 196 (3): 1012-1018.

[86] Iwan Agnieszka, Caballero-Briones F, Malinowski Marek, et al. Graphene oxide influence on selected properties of polymer fuel cells based on Nafion [J]. International Journal of Hydrogen Energy, **2017**, 42 (22): 15359-15369.

[87] He Qinggang, Li Qing, Khene Samson, et al. High-Loading Cobalt Oxide Coupled with Nitrogen-Doped Graphene for Oxygen Reduction in Anion-Exchange-Membrane Alkaline Fuel Cells [J]. Journal of Physical Chemistry C, **2013**, 117 (17): 8697-8707.

[88] Wen Qing, Wang Shaoyun, Yan Jun, et al. MnO$_2$-graphene hybrid as an alternative cathodic catalyst to platinum in microbial fuel cells [J]. Journal of Power Sources, **2012**, 216 (216): 187-191.

[89] Yan Xiao-Yan, Tong Xili, Zhang Yue-Fei, et al. Cuprous oxide nanoparticles dispersed on reduced graphene oxide as an efficient electrocatalyst for oxygen reduction reaction [J]. Chemical Communications,

2012, 48 (13): 1892-1894.

[90] Sun Meng, Liu Huijuan, Liu Yang, et al. Graphene-based transition metal oxide nanocomposites for the oxygen reduction reaction [J]. Nanoscale, **2015**, 7 (4): 1250-1269.

[91] Wang Yun, Chen Ken S., Mishler Jeffrey, et al. A review of polymer electrolyte membrane fuel cells: Technology, applications, and needs on fundamental research [J]. Applied Energy, **2011**, 88 (4): 981-1007.

[92] Choi Bong, Hong Jinkee, Chul Park Young, et al. Innovative polymer nanocomposite electrolytes: nanoscale manipulation of ion channels by functionalized graphenes [J]. Acs Nano, **2011**, 5 (6): 5167-5174.

[93] Bonaccorso Francesco, Colombo Luigi, Yu Guihua, et al. 2D materials. Graphene, related two-dimensional crystals, and hybrid systems for energy conversion and storage [J]. Science, **2015**, 347 (6217): 1246501.

[94] Yin Zongyou, Zhu Jixin, He Qiyuan, et al. Graphene-Based Materials for Solar Cell Applications [J]. Advanced Energy Materials, **2014**, 4 (1): 1300574.

[95] Nair Rahul Raveendran, Blake Peter, Grigorenko Alexander N, et al. Fine structure constant defines visual transparency of graphene [J]. Science, **2008**, 320 (5881): 1308-1308.

[96] Üregen Nurhan, Pehlivanoğlu Kübra, Özdemir Yağmur, et al. Development of polybenzimidazole/graphene oxide composite membranes for high temperature PEM fuel cells [J]. International Journal of Hydrogen Energy, **2017**, 42 (4): 2636-2647.

[97] Liu Dianyi, Li Yan, Zhao Shuli, et al. Single-layer Graphene Sheets as Counter Electrodes for Fiber-Shaped Polymer Solar Cells [J]. Rsc Advances, **2013**, 3 (33): 13720-13727.

[98] Rana Kuldeep, Singh Jyoti, Ahn Jong Hyun. A graphene-based transparent electrode for use in flexible optoelectronic devices [J]. Journal of Materials Chemistry C, **2014**, 2 (15): 2646-2656.

[99] Xu Yanfei, Long Guankui, Huang Lu, et al. Polymer photovoltaic devices with transparent graphene electrodes produced by spin-casting [J]. Carbon, **2010**, 48 (11): 3308-3311.

[100] Zhang Qian, Wan Xiangjian, Xing Fei, et al. Solution-processable graphene mesh transparent electrodes for organic solar cells [J]. Nano Research, **2013**, 6 (7): 478-484.

[101] Grätzel Michael. Photoelectrochemical cells [J]. nature, **2001**, 414 (6861): 338-344.

[102] Burschka Julian, Pellet Norman, Moon Soo-Jin, et al. Sequential deposition as a route to high-performance perovskite-sensitized solar cells [J]. Nature, **2013**, 499 (7458): 316-319.

[103] Hardin Brian E., Snaith Henry J., Mcgehee Michael D. The renaissance of dye-sensitized solar cells [J]. Nature Photonics, **2012**, 6 (6): 162-169.

[104] Wang Xuan, Zhi Linjie, Müllen Klaus. Transparen conductive graphene electrodes fordye-sensitized solar cells [J]. Nano Letters, **2008**, 8 (8): 323-327.

[105] Kim Sung Ryong, Parvez Md. Khaled, Chhowalla Manish. UV-reduction of graphene oxide and its application as an interfacial layer to reduce the back-transport reactions in dye-sensitized solar cells [J]. Chemical Physics Letters, **2009**, 483 (1): 124-127.

[106] Sun Shengrui, Gao Lian, Liu Yangqiao. Enhanced dye-sensitized solar cell using graphene-TiO_2 photoanode prepared by heterogeneous coagulation [J]. Applied Physics Letters, **2010**, 96 (8): 083113.

[107] Hong Wenjing, Xu Yuxi, Lu Gewu, et al. Transparent graphene/PEDOT-PSS composite films as counter electrodes of dye-sensitized solar cells [J]. Electrochemistry Communications, **2008**, 10 (10): 1555-1558.

第7章

氧化石墨烯在生物医学领域中的应用

如前所述，GO 具有 sp^2（C=O、C=C 等）和 sp^3（C—C、C—O—C、C—OH 等）杂化结构，表面带有大量的羟基、羧基和环氧基等含氧官能团。这些含氧官能团丰富了其表面活性，赋予了 GO 更多有趣的理化和生物学特性：a. 良好的亲水性，可以很好地分散在水中；b. 具有较大的比表面积（石墨烯理论值为 $2630m^2/g$），使得 GO 原则具有较好的载药能力；c. 同时含有疏水性的平面与亲水性的边缘，疏水性药物和染料可通过 π-π 堆积等对 GO 进行非共价修饰而负载，而含氧官能团等亲水性边缘可为功能化修饰提供活性位点；d. 光学性质及光热转换能力，使 GO 不仅能实现活体细胞的生物成像，还能在活体水平上实现光热治疗；e. 优异的机械性能，可以改善生物支架材料的空隙结构和机械性能，包括抗压强度和抗弯曲强度，加强支架的生物活性。基于这些特性，GO 已被广泛应用于生物医学和复合材料等研究领域，尤其在药物载体、生物成像、肿瘤的光热治疗及组织再造工程等方面表现出了巨大的应用潜力。

本章主要介绍 GO 的生物毒性及其机制，以及 GO 在肿瘤治疗、药物载体、生物成像、生物传感器和组织工程等方面的应用。

7.1 氧化石墨烯的生物毒性及其机制

GO 在生理条件中一般带有负电荷，通过对 GO 的修饰可以改变电荷的多少，甚至使其带上正电荷，如利用聚合物或树枝状大分子等聚阳离子试剂可以使石墨烯带有正电荷。在细胞中，GO 可能会与疏水性的、带正电荷或带负电荷的物质进行相互作用，如细胞膜、蛋白质和核酸等，因此会诱导 GO 产生毒性。在本节中，我们主要探讨 GO 在细胞（即体外）和体内试验中产生的已知毒性效应，以及产生毒性的可能原因。

石墨烯类材料主要参数包括：a.层数；b.横向尺寸；c.化学组成，如图 7.1 所示[1]。根据所采用的石墨原料片径大小、纯度高低等以及合成 GO 的方法不同，所得到 GO 片的大小、片层厚度、氧化程度（含氧量）、表面电荷和表面所带官能团等不同。GO 的生物毒性除了有浓度依赖性，还会因 GO 原料的不同而呈现出毒性数据的多样性，甚至结论相互矛盾[2~9]。此外，GO 可能与毒性测试中的试剂相互作用，从而影响细胞活性试验数据的有效性，使其产生假阳性结果。如：Macosko 等[10]人的研究发现，在细胞活性试验中利用四甲基偶氮唑盐（Methylthio Tetrazole，简称 MTT）试剂与 GO 作用后会自发生成 MTT 甲臜，而 MTT 甲臜是无毒性的，同时发现人皮肤成纤维细胞线粒体活性不受 GO 浓度的影响，因此不能体现出 GO 的细胞毒性；而当利用另一种水溶性的四唑基试剂——WST-8（台酚蓝除外）进行实验的结果与 MTT 法的结果相反，并且活性氧（Reactive Oxygen Spe-

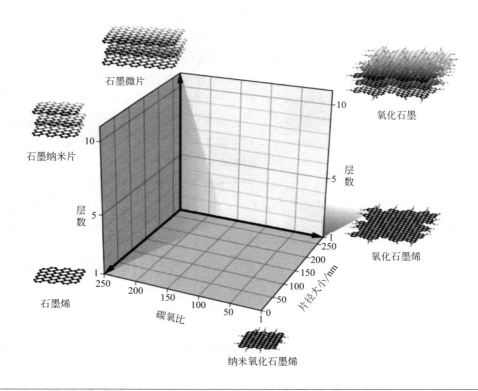

图 7.1 不同层数、横向尺寸和化学组成（碳氧比例）对石墨烯材料结构的影响[1]

cies，简称 ROS）数据表明聚集的 GO 对人皮肤成纤维细胞更具细胞毒性，可对活细胞和死细胞的数量进行精确的评估。

在细胞中，GO 的毒性与其片径、氧化程度和浓度等有关。GO 毒性效应主要有：a. 当 GO 的浓度小于 $20\mu g/mL$ 时，GO 对人成纤维细胞没有体现出明显的毒性，而当 GO 的浓度大于等于 $50\mu g/mL$ 时，就会降低人成纤维细胞的黏附，导致细胞死亡[11]。实验表明，当人肺癌（A549）细胞系中加入 $200\mu g/mL$ 的 GO 时，其 ROS 的数量是未加入 GO 的细胞系的 3.9 倍[12]。b. 一定浓度时，GO 片的大小会影响红细胞的溶血反应。如：当 GO 浓度为 $25\mu g/mL$，片径为（342 ± 17）nm 时，红细胞的溶血率在 70% 以上；当片径为（672 ± 13）nm 及（765 ± 19）nm 时，红细胞的溶血率小于 50%，甚至低于 20%；c. 一定浓度时，GO 毒性还与氧化的程度有关。当 GO 浓度为 $12.5\mu g/mL$，人皮肤成纤维细胞在（672 ± 13）nm 的 GO 溶液里的存活率约为 95%，高出在还原后石墨烯溶液中的存活率（约 83%）[10]。

GO 的体内毒性还取决于所选定的实验参数。Guo 课题组[11] 通过鼠尾静脉注射方式给老鼠注射利用改进的 Hummers' 方法制备的 GO 溶液，当使用的 GO 剂量约为 $14mg/kg$ 时，在老鼠身上观察到了 GO 的慢性毒性，有 4/9 的老鼠死亡和形成了肺肉芽肿瘤。另外对哺乳期幼鼠的实验发现，幼鼠口腔持续吸入 21 天浓度为 $0.5mg/mL$ 的 GO 溶液时，会影响幼鼠的发育，使其反应迟钝[13]。此外，Li 等[14] 人将片径约为 308nm、浓度为 $2mg/mL$ 的 GO 溶液通过气管注入的方式注入老鼠体内，注入 GO 剂量为 $10mg/kg$，三个月后发现纳米级 GO 残留在老鼠的肺里，导致老鼠急性肺损伤和慢性肺纤维化。

如同很多其他纳米材料，目前对于 GO 对细胞毒性作用的研究还没有形成统一认识。但部分 GO 生物毒性的机理已经相对明确，通常归因于 GO 与细胞膜表面直接接触从而造成细胞的破坏或氧化内部代谢物质等。为了更加深入地了解 GO 的性质，并根据其特征设计出更好的生物材料，科学家研究了 GO 的毒性机理并发展了破坏细胞膜和氧化应激等理论。下面内容对几种理论分别作相应介绍。

7.1.1 破坏细胞膜

石墨烯可通过片层的边缘切割摄取机制进入细胞膜[15]，导致细胞膜结构破坏，从而产生毒性；而 GO 及其衍生物也具有类石墨烯的边缘片段，边缘片段的范围取决于 C/O 摩尔比以及其他因素。GO 对细胞膜的影响主要取决于细胞的种类，目前针对这一机理，研究较多的细胞种类有人肺泡腺 A549 细胞[12]、人视网膜色素上皮细胞（ARPE-19）、海拉细胞、乳癌细胞（SKBR3 和 MCF-7）以及红细胞等。另外，细菌的细胞膜对石墨烯基材料也具有敏感性。

由于 GO 表面具有较高的亲和力，蛋白质可以吸附在 GO 表面，因此在生物液体中可以通过蛋白质来调节 GO 与细胞膜的相互作用。例如，血液中存在着大量的血清蛋白，可能会潜在影响 GO 的毒性。Duan 等[16] 人利用电子显微镜技术观察到牛血清蛋白可以降低 GO 对细胞膜的渗透性，抑制了 GO 对细胞膜的破坏，同时降低了 GO 的细胞毒性。基于分子动力研究分析，他们推断可能是由于 GO-蛋白质之间的作用削弱了 GO-磷脂之间的相互作用。与此同时，GO 对人血清蛋白的影响也被其他科研工作者所发现，特别是观察到了 GO 可以抑制人血清蛋白与胆红素之间的作用。可以认为 GO 与血清蛋白之间是相互影响的。

7.1.2 氧化应激

氧化应激是指体内氧化与抗氧化作用失衡，氧化倾向加强，导致中性粒细胞炎性浸润，蛋白酶分泌增加，产生大量氧化中间产物，即 ROS。大量的实验研究已经确认细胞经不同浓度的 GO 处理后，都会增加细胞中 ROS 的量。而 ROS 的量可以通过商业化的无色染料染色后利用流式细胞仪或荧光显微镜检测到。氧化应激是自由基在体内产生的一种负面作用，并被认为是导致衰老和疾病的一个重要因素。

氧化应激反应不仅与 GO 的浓度[17,18]有关，还与 GO 的氧化程度[19]有关。将蠕虫分别置于 $10\mu g/mL$ 和 $20\mu g/mL$ 的 PLL-PEG 修饰的 GO 溶液中，GO 会引起蠕虫细胞内活性氧的积累，其活性氧分别增加 59.2% 和 75.3%[18]。另外在细胞中，低氧化程度的 GO 可以促进低毒性的 H_2O_2 转变成高毒性的 HO-自由基，因此低氧化程度的 GO 更容易刺激 ROS 的产生。理论模拟显示，GO 片上的羧基和平面区域都可以改变 H_2O_2 还原反应的能量[18]。

7.1.3 其他因素

研究中发现，除了增加 ROS 外，GO 还可能会改变蛋白质的二级结构。Ding 等[20]人的研究发现，经 GO 处理过的 T 淋巴细胞可以导致 ROS 的增加、DNA 的损坏、细胞凋亡和抑制 T 淋巴细胞的免疫性，他们分析认为是由于 GO 直接与蛋白质受体进行作用，抑制了其配体结合能力。

此外，研究还发现 GO 的毒性还可能影响细胞的基因表达。如，Wu 等[21]人将 C. elegans 暴露于浓度为 100mg/mL 的 GO 溶液中，由于 GO 的毒性及其在体内迁移，导致其体内发现了几个基因的突变，包括 hsp-16.48、gas-1、sod-2、sod-3、aak-2 以及 isp-1 和 clk-1。

7.2 氧化石墨烯在生物医学领域中的应用

GO 在生理学环境下容易发生聚集，会影响其在细胞中的生物相容性和反应活性的能力，因此需要对 GO 进行功能化修饰来解决其容易团聚的问题，以更好地应用于载药、抗菌、生物成像等领域。目前功能化修饰的方式主要有以下几种。

① 共价键修饰。GO 表面丰富的含氧官能团（羟基、羧基、环氧基），可与多

种亲水性大分子通过酯键、酰胺键等共价键连接完成功能化，改善其稳定性、生物相容性等。常见的大分子有聚乙二醇、聚赖氨酸、聚丙烯和聚醚酰亚胺等；

② 非共价键修饰[22~24]。GO 片层内具有大 π 键，能够通过非共价 π-π 作用与芳香类化合物相互结合，不同种类的生物分子也可以通过氢键作用、范德华力和疏水作用等非共价作用力与 GO 结构中的 sp^2 杂化部分结合完成功能化修饰，如：生物小分子、生物大分子（核酸等）以及非生物活性化合物（药物和荧光染料等）；

③ 无机纳米颗粒修饰。利用无机纳米粒子对 GO 进行功能化修饰，如金、银、铂等金属，以及二氧化钛、氧化锌、四氧化三铁等金属氧化物，被无机纳米粒子修饰的 GO 也被广泛地应用到生物医药领域。

此外，GO 在细胞实验和体外实验中也体现出了细胞膜的渗透性和低生物毒性。由于 GO 表面具有较大的比表面积和反应活性，GO 在生物医学方面的应用具有一定潜在优势。

7.2.1 作为药物载体的应用

GO 作为一种新型的药物载体材料，以其良好的生物相容性、较高的载药率、靶向给药等方面得到广泛的关注。作为递送药物的载体，GO 不仅可以负载小分子药物，也可以与抗体、DNA、蛋白质等大分子结合。普通的有机药物很多都含有 π 结构导致这些药物的水溶性都非常差，而 GO 具有较好的亲水性，因此可以借助分散性较好的 GO 基材料来解决这个问题，即将上述药物负载到 GO 基材料上，形成 GO-药物混合物材料。这对改善难溶性药物的水溶性、降低药物不良反应以及提高药物稳定性和生物利用度等方面具有重要的意义。

（1）化学疗法药物载体

阿霉素（Doxorubicin，简称 DOX）是一种抗肿瘤抗生素，可抑制 RNA 和 DNA 的合成，抗瘤谱较广，对多种肿瘤均有作用，但对机体可产生广泛的生物化学效应，具有强烈的细胞毒性作用。目前有研究表明[25~28]：GO 基材料可以作为 DOX 及其衍生物的药物载体。由于 DOX 具有芳香环结构，可以与 GO 表面的 sp^2 杂化区域通过 π-π 堆积作用形成 GO-DOX 混合物，而且这种 π-π 堆积作用非常强，不需要添加其他共价修饰就可以使 DOX 负载到 GO 表面上。同时 DOX 又是一种荧光分子，因此当 GO-DOX 进入细胞后可以通过流式细胞仪或荧光显微镜进行监测，通过光学成像观察药物受体位置和药物反应，为药物疗效评价提供直接信息。

Yin 等[29] 利用叶酸修饰 GO，通过对 GO 溶液超声和低温下（4℃）长时间负载 DOX（记作 HGO-DOX），显示 GO 对 DOX 的负载率可高达 49%。该研究把 HGO-DOX 加入到 pH 为 5.5 的磷酸缓冲溶液中，并于 37℃恒温振荡箱中考察药物的体外释放。结果表明，用透明质酸酶处理过的 HGO-DOX，当释放时间达到 16 小时后，药物累积释放量为 45%，具有缓释性。Xing 等[30] 人利用 PEI 修饰 GO 后并通过静电吸附作用与康宁酸改性的聚丙烯胺（记作 PAH-Cit）进行结合，

DOX通过共价键与PAH-Cit结合而负载到GO上，DOX的负载量随着DOX初始浓度的增加而成线性增加；当初始浓度为0.32mg/mL时，DOX的负载量可以达到0.294mg/mg。同时利用不同pH条件在37℃时研究DOX的释放实验，结果发现随着pH由7.4降至5.0，DOX的释放量由28.75%增加至81.25%。Wei等[31]人利用转铁蛋白（Transferrin，简称Tf）和双氢青蒿素（Dihydroartemisinin，简称DHA）修饰GO，得到DHA-GO-Tf，然后通过小鼠尾部静脉注射方式分别将剂量为0.2mg/kg的DHA和0.2mg/kg DHA-GO-Tf在30天里每隔48小时完成注射，并每2天观察肿瘤肿块的大小。实验结果显示：在30天内，注射相同剂量的药物，GO负载DHA和Tf实验组的小鼠肿瘤已经痊愈，15天后肿瘤肿块就已经开始结痂，随着时间的延长，结痂开始脱落，而纯DHA实验组的老鼠肿块还没有痊愈。

除此之外，GO基材料还可以负载含有金属颗粒的抗癌药物以及光敏剂、具有生物活性的核酸适配体[32]、黄酮类槲皮素[33]、叶酸[34]及其他药物[35,36]。

（2）光热治疗药物载体

光热治疗法是基于具有较高光热转换效率的材料，利用靶向性识别技术聚集在肿瘤组织或伤口中的细菌附近，并在外部光源（一般是近红外光）的照射下将光能转化为热能来杀死癌细胞、活细菌的一种治疗方法。近年来，由于高渗透性和滞留效应（Enhanced Permeability and Retention Effect，简称EPR），GO被当成纳米载体用到癌症的靶向治疗中，包括光热治疗和光动力治疗，以达到改善药物对细胞膜的渗透性以及聚集在肿瘤细胞部位的目的[7~9,37,38]。在光热治疗法中，为了防止健康组织受到热损伤，有两个光学波段需要避免照射到健康组织上，一个是700～980nm波段，另一个是1000～1400nm波段，也被称为两个生物学窗口。因为这两个波段的光可以通过几厘米厚的人体组织[39]。单层GO具有良好的水溶性、细胞膜渗透性和较好的稳定性，同时GO也可以吸收近红外波段的光[40,41]，因此非常适合用于光热治疗法中。GO对近红外光的吸收率受GO尺寸大小的影响，小尺寸GO（<300nm）吸收近红外光的效率要远高于其他的材料。由于实体瘤的高通透性和滞留效应，会促使最佳尺寸大小的GO片集聚在肿瘤部位。此外，可以通过相应的配体修饰GO使其聚集在特定的肿瘤受体部位[42~45]。

Rong等[46]人利用聚乙二醇功能化修饰的GO（GO-PEG），通过π-π共轭作用负载上近红外荧光染料分子（CySCOOH），以此来增强GO的近红外吸收性能，并研究了GO-PEG-CySCOOH对肿瘤的光热治疗效果。实验中，分别对4T1肿瘤模型鼠注射剂量为168μg的CySCOOH，300μg的GO-PEG和300g的GO-PEG-CySCOOH，然后将老鼠置于波长为808nm，功率密度为0.5W/cm^{2}的近红外光下辐射5分钟，进行60天的光热治疗。实验结果显示，当光热治疗14天后，注射GO-PEG-CySCOOH模型鼠的肿瘤已经愈合，而其他组的肿瘤生长未见明显的变化；经过60天的光热治疗后，模型鼠的肿瘤没有复发，认为GO-PEG-CySCOOH

在光热治疗肿瘤实验中具有较好的疗效。Li 等[47]人利用镧掺杂的纳米颗粒 $NaLuF_4$：Er^{3+}、Yb^{3+} 修饰纳米 GO（UCNP@NGO），先将 4T1 乳腺肿瘤细胞置于加入纳米 GO 和 UCNP@NGO 的 10%磷酸缓冲盐溶液（PBS）里在 37℃下培养 24 小时，研究纳米 GO 和 UCNP@NGO 的细胞毒性，其中纳米 GO 和 UCNP@NGO 的浓度分别为 $0\mu g/mL$、$50\mu g/mL$、$100\mu g/mL$、$200\mu g/mL$ 和 $400\mu g/mL$。实验结果表明，4T1 乳腺肿瘤细胞在 $400\mu g/mL$ 的纳米 GO 和 UCNP@NGO 培养液中，细胞存活率高达 90%以上，两者都未显示出明显的毒性。然后又分别对带有 4T1 乳腺肿瘤的老鼠注射 $100\mu L$ 的 PBS、纳米 GO 和 UCNP@NGO，其中纳米 GO 和 UCNP@NGO 的剂量分别为 $20\mu g$。并用 808nm 的近红外光辐射 5 分钟，激光功率为 $1W/cm^2$。近红外光辐射的同时利用红外相机进行监测，结果发现辐射 5 分钟后，注射 UCNP@NGO 的 4T1 乳腺肿瘤部位的温度时升至约 63℃，比注射纳米 GO 的高出 10℃。光热实验结果显示，纳米 GO 与 UCNP@NGO 在近红外光的辐射下都对肿瘤细胞具有抑制作用。

（3）联合疗法药物载体

从肿瘤治疗的应用角度，GO 本身既有光热效应，同时又能高效负载化疗药物和生物靶向分子，使其在肿瘤联合治疗方面的优势也很突出。相比于传统的化学疗法，采用 GO 的联合疗法可以改善在传统的化学疗法中出现的一些不良反应，同时还可以达到更好的协同效应。Zhang 等[48]人通过共价修饰方法将聚乙二醇（PEG）和 DOX 修饰到纳米 GO 上，得到纳米 GO-PEG-DOX 材料，并利用肿瘤移植模型鼠试验来研究纳米 GO-PEG-DOX 材料的抗肿瘤活性。实验采用的激光功率密度为 $2W/cm^2$，照射斑点为（6×8）mm，时间为 5 分钟。实验结果显示：经过 30 天的治疗，肿瘤完全被毁坏。另外，纳米 GO-PEG-DOX 材料组小鼠的不良反应明显低于 DOX 组。Li 等[49]利用三嵌段共聚物 F127 对 GO 进行功能化修饰，通过 π-π 作用、疏水作用和氢键作用将 DOX 负载到 GO 上，得到 F127-GO-DOX，然后对老鼠分别注射 $100\mu L$ 的生理盐水、含 1mg/mL F127-GO 的生理盐水和含 1mg/mL F127-GO-DOX 的生理盐水，并采用功率密度为 $0.5W/cm^2$、808nm 波长的近红外光进行辐射，辐射时间为 10 分钟。实验结果显示，经过 14 天的辐射治疗后，注射 F127-GO-DOX 的肿瘤组织没有明显的生长，且已经变成黑色的结痂。以上实验结果都说明了 DOX 与光热治疗法具有较强的协同抗癌效应，对改善目前治疗癌症的方法具有指导意义。

7.2.2 抗菌

人的免疫系统出现问题或者病人的伤口，如裂伤、割伤、刺伤、盲管伤、穿通伤等，都很容易受到细菌感染，导致处于危险之中。另外，伤口的程度与伤口的大小、深浅、污染程度等密切相关，将会引起不同类型的感染。感染程度越大，治疗时间越长、难度大且不易愈合。因此，开发新型的抗菌药物是科研工作者努力的方向。

GO 具有疏水性中间片层与亲水性边缘结构，因而具有抗菌特性。GO 的抗菌活性主要有以下几种机制：a. 机械破坏，包括物理穿刺或者切割；b. 氧化应激引发的细菌/膜物质破坏；c. 包覆导致的跨膜运输阻滞和（或）细菌生长阻遏；d. 磷脂分子抽提理论。GO 作用于细菌膜表面的杀菌机制，主要是 GO 与起始分子反应（Molecular Initiating Events，简称 MIEs）[50] 的作用，包括 GO 表面活性引发的磷脂过氧化，GO 片层结构对细菌膜的嵌入、包裹以及磷脂分子的提取，GO 表面催化引发的活性自由基等。另外，不同尺寸的 GO 在上述不同的抗菌机制中对抗菌的影响也是不同的，机械破坏和磷脂分子抽提理论表明尺寸越大的 GO，能表现出更好的抗菌能力，而氧化应激理论则认为 GO 尺寸越小，其抗菌效果越好。

GO 的抗菌活性一种可能的原因为：片状结构的锋利边缘引起的膜应力，通过刺入细胞体导致物理破坏，导致细菌膜完整性的丧失和生物物质的泄漏。对 GO 基材料的少数细胞毒性研究表明，一些抗菌活性与"物理"效应和"化学"效应的协同作用有关[51]。如图 7.2(a) 所示，将 GO 和 RGO 组装成纸状薄膜来考察对大肠杆菌生长的最小抑制，以显示其最小的细胞毒性[52]。当细菌直接接触石墨烯材料时，强烈的物理作用可能对细胞膜造成物理损伤并导致细胞内物质的释放，而 GO 膜则通过化学方法增加细胞氧化应激反应和活性自由基，这个过程可能破坏了特定的微生物过程[53]。将 MOS_2 与 GO 复合以形成纳米复合膜，对革兰氏阴性菌大肠杆菌具有更好的抗微生物特性，谷胱甘肽氧化能力和部分电导率均有所增加[54]。此外，研究人员还制备出一种可回收纳米复合材料，即 GO（GOIONP-Ag）表面上的氧化铁纳米颗粒（IONP）和银纳米颗粒（AgNPs）来作为多功能抗菌材料[55]。图 7.2(b) 展示了采用抑制滤波器通过真空过滤法将 GO 悬浮液制备成高度褶皱的 GO 膜，表现出优异的抗菌性能，表明临界形状诱导的结构对抗菌性能有一定影响。另外，研究还证明此 GO 膜过滤器对于压滤和光照射去除细菌是无活性的。科研人员还通过实验评估了 GO 纳米片的化学状态和缺陷，并验证了其产生抗菌性质的自由基（活性氧物质）。从 GO 释放的碳自由基具有抗生素抗性细菌灭活，揭示了 GO 膜中实现抗菌作用的主要原理，参照图 7.2 所示。石墨烯纳米片显示出对大肠杆菌和鼠伤寒沙门氏菌（$1\mu g/mL$）、粪肠球菌（$8\mu g/mL$）和枯草芽孢杆菌（$4\mu g/mL$）的最小抑制浓度，具有与商业细菌抗生素相当的抗菌活性[56]。

Khan 等[59] 人采用 Hummers' 法制备出 GO 材料，并利用光热治疗法将其应用到处理伤口的细菌感染实验中。该实验也是少数在体内研究常规化学修饰 GO 的生物学活性的例子之一。实验中，研究人员统计了所研究的健康小白鼠，每个小白鼠都有三处伤口，并让它们全部感染上金黄色酿脓葡萄球菌。三处伤口处理如下：一处不处理，作为对照组；另一处利用 Nd：YAG 激光器照射，照射光波长为 1064nm，每天照 3 次，持续 12 天；第三处伤口用 $15\mu L$ 浓度为 0.2mg/mL 的 GO 溶液处理后，再利用 Nd：YAG 激光器照射，照射光波长为 1064nm，每天照 3 次，

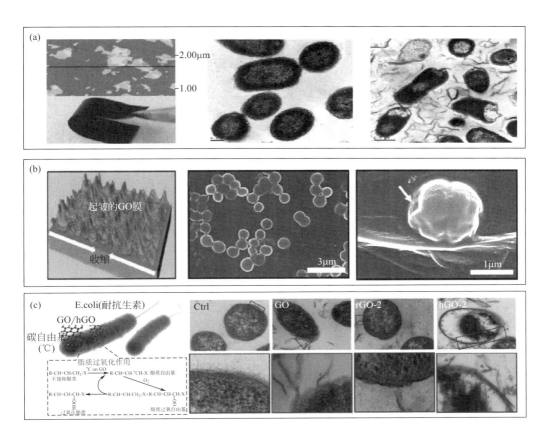

图 7.2 GO 膜的抗菌应用

(a) 通过真空过滤制备的独立且柔韧的 GO 和 RGO 纸及其对大肠杆菌 37℃的抗菌活性,持续 2h[52];(b) GO 膜表面的皱纹结构及其通过各种细胞相互作用相应地直接影响细菌活性[57];(c) 通过碳自由基的 GO 膜失活和 TEM 图像的示意图,显示 GO 膜和大肠杆菌之间的相互作用[58]

持续 12 天。结果显示:经过 GO 和光照射处理的伤口,其康复速度最快。这些数据表明,相对于抗生素,利用 GO 材料结合光热治疗法是一种更有效的治疗细菌感染的方法。

除此之外,寻找一种生物相容性好并且具有抗菌作用的材料可以很大程度上改善治疗效果。大量研究表明,GO 具有良好的抗菌作用[52,60~62]。大肠杆菌是人和许多动物肠道中最主要且数量最多的一种细菌,为埃希氏菌属代表菌。GO 纳米溶液对大肠杆菌具有较好的抑制性,其抗菌性主要源于其对大肠杆菌细胞膜的破坏。然而,目前对于 GO 材料的抗菌能力仍然缺乏系统的研究,因而深入研究其抗菌机理,以期更好去使用这种材料显得很必要。

7.2.3 生物成像

生物光学成像是指利用光学的探测手段结合探测分子对细胞或者组织甚至生物体进行成像，来获得其中的生物学信息的方法。如果把生物光学成像限定在可见光和近红外光范围，依据探测方式的不同，生物光学成像可分为荧光成像、生物发光成像、光声成像、核素成像等。生物光学成像能够无损实时动态监测被标记细胞在活体小动物体内的活动及反应，在肿瘤检测、基因表达、蛋白质分子检测、药物受体定位、药物筛选和药物疗效评价等方面具有很大的应用潜力。

由于较低的毒性和良好的生物相容性，石墨烯材料在细胞成像方面引起了关注。石墨烯及其衍生物本身具有的特殊平面结构和光学性质，或者经过荧光染料分子标记之后，可用于体外细胞与活体光学成像[63~66]，使其可能应用于肿瘤显像和治疗。Sun等[67]利用纳米尺寸的聚乙二醇功能化GO（记作GO-PEG）的近红外发光性质用于细胞成像。他们将抗体利妥昔单抗（anti-CD20）与纳米GO-PEG共价结合形成纳米GO-PEG-anti-CD20，然后将纳米GO-PEG和纳米GO-PEG-anti-CD20与B细胞或T细胞在培养液中4℃培养1小时，培养液中纳米GO-PEG的浓度大约为0.7mg/mL，结果发现B细胞淋巴瘤具有强荧光，而T淋巴母细胞的荧光强度则很弱。另外，通过对GO进行80℃热处理17天后，再利用200W的超声对GO溶液处理2小时，得到的GO在紫外光（266~340nm）的照射下显示出蓝色荧光。然后在小鼠结直肠癌细胞系CT26培养液中加入5μg/mL的GO溶液后继续培养24小时，利用AlexaFluor® 488Phalloidin染料分子对细胞进行染色，结果发现经过上述方法处理的GO在细胞中显示出绿色和蓝色荧光，而没有处理的GO只显示出绿色荧光[68]。由此看出GO在细胞成像和相关领域具有潜在的应用。

另一方面，纳米GO的近红外光容易受到生物组织的干扰，会限制其动物活体成像，对于这种情况可以利用荧光染料分子对GO进行标记来用于光学成像。如利用含硅的酞菁染料对GO进行染色后（记作SiPc-GO），其近红外光的吸收率是纯GO的三倍，将Hela癌细胞在10μg/mL的SiPc-GO溶液后继续培养24h，实验结果显示SiPc-GO可以进入细胞内，使细胞质内具有荧光性[69]。

另外，光学成像往往会面临荧光淬灭与组织穿透深度限制的问题，但是核素成像，包括正电子发射断层显像（Positron Emission Tomography，简称PET）和单光子发射断层显像（Single-Photon Emission Tomography，简称SPECT）能够克服上述问题。核素成像需要将放射活性化学物质注射到体内，采用放射性核素示踪的间接检测技术，是目前灵敏度最高和易于定量的成像模式。因此有研究利用核素标记抗体后再与GO材料进行结合[63~65]，希望克服荧光淬灭与组织穿透深度限制的问题。Hong等[63]人利用^{64}Cu标记抗体TRC105共轭结合纳米尺寸的GO-PEG，并进行4T1肿瘤模型的PET成像，结果发现可以有效靶向肿瘤组织。此外，

Cornelissen 等[64] 人利用放射性 ^{111}In 标记抗体曲妥单抗（Tz）修饰的纳米 GO，进行肿瘤模型鼠的 SPECT 成像，并与封闭组以及非 HER2 受体特异性的 IgG 蛋白修饰的 ^{111}In-GO-IgG 和 Tz 抗体本身 ^{111}In-Tz 进行比较。通过 SPECT 成像可清晰观察到肿瘤部位，^{111}In-GO-Tz 可以被肿瘤组织高度吸收，并且 ^{111}In-GO-Tz 与 ^{111}In-GO-IgG 和 ^{111}In-Tz 相比，具有更好的肿瘤成像特性。

由于核素成像的优势，值得进一步开展核素标记的 GO 体系及其在 PET 和 SPECT 方面的研究。此外，通过一定功能化修饰，包括引入显像信号和治疗药物，得到的 GO 体系既可以成像，又能治疗，将在显像指导的肿瘤治疗方面拥有潜力。除此之外，GO 的固有光声特性还能实现光声成像，GO 在光声成像方面也表现出一定的应用潜力[66]。在不久的将来，GO 基材料的生物成像有望真正地应用于医学领域，为临床诊断提供直接有效的信息。

7.2.4 生物传感器

GO 的比表面积很大，而厚度只有几纳米，具有两亲性，表面的各种官能团使其可与生物分子直接相互作用，易于化学修饰，同时具有良好的生物相容性，超薄的 GO 纳米片很容易组装成纸片或直接在基材上进行加工。另外，GO 可以通过荧光能量共振转移和非辐射偶极-偶极相互作用，可以有效淬灭荧光体（染料分子、量子点及上转换纳米材料）的荧光。这些特点都使 GO 成为制作传感器良好材料选择[62,70,71]。Arben 等人的研究发现，将 CdSe/ZnS 量子点作为荧光供体，石墨、碳纤维、碳纳米管和 GO 作为荧光受体，对 CdSe/ZnS 量子点的荧光淬灭效率分别为 66%±17%、74%±7%、71%±1% 和 97%±1%。与其他碳材料相比，GO 具有更好的荧光猝灭效果[72]。

利用上述特性，可以构建以下几种 GO 型传感器。a. 将 GO 作为荧光共振能量转移的受体，构建荧光共振能量转移型 GO 生物传感器，用于检测各种生物分子。b. 将一些抗体键合在 GO 表面，构建成抗体型 GO 传感器。通常是将 GO 作为荧光共振能量转移或化学发光共振能量转移的受体，以此来检测抗原物质；或者利用 GO 比表面积较大能结合更多抗体的特点，将检测信号进行进一步放大。c. 构建多肽型 GO 传感器。因为 GO 是一种边缘含有亲水基团（—COOH，—OH 及其他含氧基团）而基底具有高疏水性的两性物质，当多肽与 GO 孵育时，多肽的芳环和其他疏水性残基与 GO 的疏水性基底堆积，二者部分残基之间同时也会存在静电作用，这样多肽组装在 GO 上形成了多肽型 GO 传感器。当多肽被荧光基团标记时，二者之间发生荧光共振能量转移后，GO 使荧光发生猝灭。

GO 传感器在生物分子检测方面具有高灵敏度、高特异性，可以快速地检测出 DNA[65,66]、蛋白质、脱氧核糖核酸、多肽[62]、罗丹明和多巴胺[73] 等。GO 基生物传感器同时也适用于其他肿瘤细胞表面标记评价系统，为人们提供了一种简单高效、灵敏快速的检测方法。

7.2.5 组织工程

目前医学界面临的一个棘手难题是对大面积骨组织缺损的修复。其中,干细胞治疗可能是一种很有前途的解决方案,但是在干细胞的移植过程中,需要可促进和增强细胞成活、附着、迁移和分化并有着良好生物相容性的支架材料。研究表明,GO 具有良好的生物相容性及较低的细胞毒性,可促进成纤维细胞、成骨细胞和间充质干细胞(Mesenchymal Stem Cells,简称 MSC)的增殖和分化[74],同时 GO 还可以促进多种干细胞的附着和生长,增强其成骨分化的能力[75,76]。因此 GO 受到了骨组织再生领域及相关领域研究人员的关注,成为组织工程研究中一种有潜力的支架材料。GO 不仅可以单独作为干细胞的载体材料,还可以加入到现有的支架材料中;GO 不仅可以加强支架材料的生物活性,同时还可以改善支架材料的空隙结构和机械性能,包括抗压强度和抗弯曲强度。GO 表面积及粗糙度较大,适合 MSC 的附着和增殖,从而可促进间充质干细胞的成骨分化,而这种作用程度与支架中加入 GO 的比例成正比[77~79]。

Zhou 等[80] 人利用电纺丝法制备出聚-3-羟基丁酸-4-羟基丁酸/GO(P34HB/GO)复合材料支架,GO 的加入减小了纤维直径,增强了支架的多孔性、亲水性、机械强度和成骨分化性能。研究人员将老鼠麻醉后通过手术移去直径为 5mm 的颅骨,再将 P34HB 和 P34HB/GO(含有 0.66% 的 GO)支架膜分别裁成直径为 5mm 的圆片,置于去除颅骨的部位,然后将骨膜、肌肉和皮肤缝合。手术后的老鼠每天注射剂量为 10000U 的盘尼西林,并持续三天。实验 4 周时,P34HB/GO 支架膜实验组的新骨密度达到 34.17%,而 P34HB 支架膜实验组基本上没有明显变化;实验 8 周后,置有 P34HB/GO 支架膜的新骨生长率为 66.77%,高于 P34HB 支架膜的 40.20%。由此,可以认为 GO 的加入可以促进成骨分化,在骨组织修复方面有一定的应用潜力。

同时,GO 的载药作用也可促进间充质干细胞的成骨分化。如用携带正电荷 NH^{3+} 的 GO($GO-NH^{3+}$)和携带负电荷 $COOH^-$ 的 GO($GOCOOH^-$)可以交替层叠得到最外层为 $GO-COOH^-$ 的 GO 载体。以此载体携带骨形态发生蛋白-2(BMP-2)和 P 物质(SP)附着到钛(Ti)种植体上,研究发现表面覆盖 GO-COOH-、携带 BMP-2 和 SP(Ti/GO-/SP/BMP-2)种植体周围的新骨生成量要明显多于 Ti/SP/BMP-2、Ti/GO-/BMP-2、Ti/GO-/SP。这说明 GO 可以同时携带 BMP-2 和 SP 到达局部并缓慢释放,增加局部 BMP-2 和 SP 的有效剂量且发挥生物活性作用[81,82]。GO 的这种双重携带传递作用可在口腔种植及骨愈合方面起到重要作用。体内羟磷灰石(Hydroxyapatite,简称 HA)是一种常用于骨组织修复的磷酸钙陶瓷类材料。在 HA 中加入 GO,可以增强其在钛板表面的附着强度;以 HA 为基底,表面覆盖 GO 的复合物(GO/HA)表现出比纯 HA 更高的抗腐蚀性能,细胞活性也更强[83,84]。

7.3 小结

GO 生物相容性的研究已经积累了一定的研究基础，但 GO 在实际应用中仍然面临很多困难和挑战。首先，GO 制备方法的多样性和生物系统的复杂性，会显著影响其在体内外的生物相容性，导致研究结果的不一致。因此，对于 GO 的生物相容性问题不能简单归纳得出结论，需要综合多方面的因素进行深入研究。其次，GO 的抗菌活性也取决于时间和 GO 的浓度，其抗菌机理需要进一步的研究。最后，GO 对机体的长期毒性以及 GO 进入细胞的机制、与细胞之间相互作用的机理、细胞/体内代谢途径等尚不清晰。这些问题关乎 GO 在生物医学领域应用中的安全问题和环境风险评价，需要研究者们不断地研究和探索。

在推动以 GO 为载体的新药进入临床试验前，势必会面临诸多挑战：a.优化 GO 的制备方法及生产工艺，使其具有可重复性，并能精确控制 GO 的尺寸和质量；b.最佳使用剂量的摸索，找到以 GO 为载体的处方疗效和毒性之间的平衡点；c.其他表面修饰剂的开发，需具有良好生物相容性且修饰后的 GO 能在短时间内被生物体清除；d.毒理学方法的进一步规范，系统阐明以 GO 为载体处方的潜在毒性；e.体内外模型的建立，全面评价 GO 处方的生物相容性，使其能更好地转化到临床。以 GO 为载体的处方在大规模工业化生产和应用时，还需考虑到对人体和环境的不利影响，是否存在可能导致潜在的风险暴露和环境污染问题，这些有待于进一步研究。

参考文献

[1] Peter Wick, Louw-Gaume Anna E., Kucki Melanie, et al. Classification framework for graphene-based materials [J]. Angewandte Chemie, **2014**, 53 (30): 7714-7718.

[2] Alberto Bianco. Graphene: safe or toxic? The two faces of the medal [J]. Angewandte Chemie International Edition, **2013**, 52 (19): 4986-4997.

[3] Toktam Nezakati, Cousins Brian G, Seifalian Alexander M. Toxicology of chemically modified graphene-based materials for medical application [J]. Archives of Toxicology, **2014**, 88 (11): 1987-2012.

[4] Jachak Ashish C., Megan Creighton, Yang Qiu, et al. Biological interactions and safety of graphene materials [J]. Mrs Bulletin, **2012**, 37 (12): 1307-1313.

[5] Guo Xiaoqing, Nan Mei. Assessment of the toxic potential of graphene family nanomaterials [J]. Journal of Food & Drug Analysis, **2014**, 22 (1): 105-115.

[6] Seabra Amedea, Paula Amauri, Lima Renata, et al. Nanotoxicity of graphene and graphene oxide [J]. Chemical Research in Toxicology, **2014**, 27 (2): 159-168.

[7] Goenka Sumit, Sant Vinayak, Sant Shilpa. Graphene-based nanomaterials for drug delivery and tissue engineering [J]. Journal of Controlled Release Official Journal of the Controlled Release Society, **2014**, 173

(1): 75-88.

[8] Vijaya Krishna K, Cécilia Ménard Moyon, Sandeep Verma, et al. Graphene-based nanomaterials for nanobiotechnology and biomedical applications [J]. Nanomedicine, **2013**, 8 (10): 1669-1688.

[9] Skoda Marta, Dudek Ilona, Jarosz Anna, et al. Graphene: one material, many possibilities—application difficulties in biological systems [J]. Journal of Nanomaterials, **2014**, 11.

[10] Liao Ken-Hsuan, Lin Yu-Shen, Macosko Christopher W., et al. Cytotoxicity of graphene oxide and graphene in human erythrocytes and skin fibroblasts [J]. Acs Appl Mater Interfaces, **2011**, 3 (7): 2607-2615.

[11] Wang Kan, Ruan Jing, Song Hua, et al. Biocompatibility of Graphene Oxide [J]. Nanoscale Research Letters, **2011**, 6 (1): 6-8.

[12] Chang Yanli, Yang Sheng Tao, Liu Jia Hui, et al. In vitro toxicity evaluation of graphene oxide on A549 cells [J]. Toxicology Letters, **2011**, 200 (3): 201-210.

[13] Fu Changhui, Liu Tianlong, Li Linlin, et al. Effects of graphene oxide on the development of offspring mice in lactation period [J]. Biomaterials, **2015**, 40 (3): 23-31.

[14] Li Bo, Yang Jianzhong, Huang Qing, et al. Biodistribution and pulmonary toxicity of intratracheally instilled graphene oxide in mice [J]. Npg Asia Materials, **2013**, 5 (5): 237-239.

[15] Li Yinfeng, Yuan Hongyan, Bussche Annette von dem, et al. Graphene microsheets enter cells through spontaneous membrane penetration at edge asperities and corner sites [J]. Proceedings of the National Academy of Sciences of the United States of America, **2013**, 110 (30): 12295-12300.

[16] Duan Guangxin, Kang Seung-Gu, Tian Xin, et al. Protein corona mitigates the cytotoxicity of graphene oxide by reducing its physical interaction with cell membrane [J]. Nanoscale, **2015**, 7 (37): 15214-15224.

[17] Wu Qiuli, Yin Li, Li Xing, et al. Contributions of altered permeability of intestinal barrier and defecation behavior to toxicity formation from graphene oxide in nematode Caenorhabditis elegans [J]. Nanoscale, **2013**, 5 (20): 9934-9943.

[18] Zhang Wendi, Wang Chi, Li Zhongjun, et al. Unraveling stress-induced toxicity properties of graphene oxide and the underlying mechanism [J]. Advanced Materials, **2012**, 24 (39): 5391-5397.

[19] Zhang Wendi, Yan Liang, Li Meng, et al. Deciphering the underlying mechanisms of oxidation-state dependent cytotoxicity of graphene oxide on mammalian cells [J]. Toxicology Letters, **2015**, 237 (2): 61-71.

[20] Ding Zhijia, Zhang Zhijun, Ma Hongwei, et al. In vitro hemocompatibility and toxic mechanism of graphene oxide on human peripheral blood T lymphocytes and serum albumin [J]. Acs Appl Mater Interfaces, **2014**, 6 (22): 19797-19807.

[21] Wu Qiuli, Zhao Yunli, Li Yiping, et al. Molecular signals regulating translocation and toxicity of graphene oxide in the nematode Caenorhabditis elegans [J]. Nanoscale, **2014**, 6 (19): 11204-11212.

[22] Shi Sixiang, Chen Feng, Ehlerding Emily, et al. Surface engineering of graphene-based nanomaterials for biomedical applications [J]. Bioconjug Chem, **2014**, 25 (9): 1609-1619.

[23] Miao Wenjun, Shim Gayong, Lee Sangbin, et al. Safety and tumor tissue accumulation of pegylated graphene oxide nanosheets for co-delivery of anticancer drug and photosensitizer [J]. Biomaterials, **2013**, 34 (13): 3402-3410.

[24] Weaver Cassandra L, Larosa Jaclyn M, Luo Xiliang, et al. Electrically controlled drug delivery from graphene oxide nanocomposite films [J]. Acs Nano, **2014**, 8 (2): 1834-1843.

[25] Zhao Xubo, Liu Lei, Li Xiaorui, et al. Biocompatible graphene oxide nanoparticle-based drug delivery

platform for tumor microenvironment-responsive triggered release of doxorubicin [J]. Langmuir the Acs Journal of Surfaces Colloids, **2014**, 30 (34): 10419-10429.

[26] Yang Xiaoying, Wang Yinsong, Huang Xin, et al. Multi-functionalized graphene oxide based anticancer drug-carrier with dual-targeting function and pH-sensitivity [J]. Journal of Materials Chemistry, **2011**, 21 (10): 3448-3454.

[27] Suryaprakash Smruthi, Li Mingqiang, Lao Yeh Hsing, et al. Graphene oxide cellular patches for mesenchymal stem cell-based cancer therapy [J]. Carbon, **2018**, 129: 863-868.

[28] Qin Yu, Wang Changyu, Jiang Yun, et al. Phosphorylcholine oligomer-grafted graphene oxide for tumor-targeting doxorubicin delivery [J]. Rsc Advances, **2017**, 7 (66): 41675-41685.

[29] Yin Tingjie, Liu Jiyong, Zhao Zekai, et al. Redox Sensitive Hyaluronic Acid-Decorated Graphene Oxide for Photothermally Controlled Tumor-Cytoplasm-Selective Rapid Drug Delivery [J]. Advanced Functional Materials, **2017**, 27 (14): 1604620.

[30] Zhou Ting, Zhou Xiaoming, and Xing Da. Controlled release of doxorubicin from graphene oxide based charge-reversal nanocarrier [J]. Biomaterials, **2014**, 35 (13): 4185-4194.

[31] Liu Lijuan, Wei Yanchun, Zhai Shaodong, et al. Dihydroartemisinin and transferrin dual-dressed nano-graphene oxide for a pH-triggered chemotherapy [J]. Biomaterials, **2015**, 62: 35-46.

[32] Wang Ying, Li Zhaohui, Hu Dehong, et al. Aptamer/graphene oxide nanocomplex for in situ molecular probing in living cells [J]. Journal of the American Chemical Society, **2016**, 132 (27): 9274-9276.

[33] Rahmanian Nazanin, Hamishehkar Hamed, Dolatabadi Jafar Ezzati Nazhad, et al. Nano graphene oxide: A novel carrier for oral delivery of flavonoids [J]. Colloids Surfaces B Biointerfaces, **2014**, 123: 331-338.

[34] Yang Xiaoying, Niu Gaoli, Cao Xiufen, et al. The preparation of functionalized graphene oxide for targeted intracellular delivery of siRNA [J]. Journal of Materials Chemistry, **2012**, 22 (14): 6649-6654.

[35] Kim Hyunwoo, Lee Duhwan, Kim Jnhwan, et al. Photothermally triggered cytosolic drug delivery via endosome disruption using a functionalized reduced graphene oxide [J]. Acs Nano, **2013**, 7 (8): 6735-6746.

[36] Ni Yongnian, Zhang Fangyuan, Kokot Serge. Graphene oxide as a nanocarrier for loading and delivery of medicinal drugs and as a biosensor for detection of serum albumin [J]. Analytica Chimica Acta, **2013**, 769 (1): 40-48.

[37] Sun Bingmei, Wu Jinrui, Cui Shaobin, et al. In situ synthesis of graphene oxide/gold nanorods theranostic hybrids for efficient tumor computed tomography imaging and photothermal therapy [J]. Nano Research, **2017**, 10 (1): 37-48.

[38] Zhang Liming, Xia Jingguang, Zhao Qinghuan, et al. Functional graphene oxide as a nanocarrier for controlled loading and targeted delivery of mixed anticancer drugs [J]. Small, **2010**, 6 (4): 537-544.

[39] Helmchen Fritjof, Denk Winfried. Deep tissue two-photon microscopy [J]. Nature Methods, **2005**, 2 (12): 932.

[40] Saxena Sumit, Tyson T, Shukla Shobha, et al. Investigation of structural and electronic properties of graphene oxide [J]. Applied Physics Letters, **2011**, 99 (1): 013104.

[41] Li Meng, Yang Xinjian, Ren Jinsong, et al. Using graphene oxide high near-infrared absorbance for photothermal treatment of Alzheimer's disease [J]. Advanced Materials, **2012**, 24 (13): 1722-1728.

[42] Tian Bo, Wang Chao, Zhang Shuai, et al. Photothermally enhanced photodynamic therapy delivered by nano-graphene oxide [J]. Acs Nano, **2011**, 5 (9): 7000-7009.

[43] Guo Miao, Huang Jie, Deng Yibin, et al. pH-Responsive Cyanine-Grafted Graphene Oxide for Fluores-

cence Resonance Energy Transfer-Enhanced Photothermal Therapy [J]. Advanced Functional Materials, **2015**, 25 (1): 59-67.

[44] Robinson Joshua T, M Tabakman Scott, Liang Yongye, et al. Ultrasmall reduced graphene oxide with high near-infrared absorbance for photothermal therapy [J]. Journal of the American Chemical Society, **2011**, 133 (17): 6825-6831.

[45] Zhang Liming, Lu Zhuoxuan, Zhao Qinghuan, et al. Enhanced chemotherapy efficacy by sequential delivery of siRNA and anticancer drugs using PEI-grafted graphene oxide [J]. Small, **2015**, 7 (4): 460-464.

[46] Rong Pengfei, Wu Jianzhen, Liu Zhiguo, et al. Fluorescence Dye Loaded Nano-graphene for Multimodal Imaging Guided Photothermal Therapy [J]. Rsc Advances, **2015**, 6 (3): 1894-1901.

[47] Li Po, Yan Yue, Chen Binglong, et al. Lanthanide-doped upconversion nanoparticles complexed with nano-oxide graphene used for upconversion fluorescence imaging and photothermal therapy [J]. Biomaterials Science, **2018**, 6 (4): 877-884.

[48] Zhang Wen, Guo Z, Huang Deqiu, et al. Synergistic effect of chemo-photothermal therapy using PEGylated graphene oxide [J]. Biomaterials, **2011**, 32 (33): 8555-8561.

[49] Li Bingxia, Zhang Luna, Zhang Zichen, et al. Physiologically stable F127-GO supramolecular hydrogel with sustained drug release characteristic for chemotherapy and photothermal therapy [J]. Rsc Advances, **2018**, 8 (3): 1693-1699.

[50] Zheng Huizhen, Ma Ronglin, Gao Meng, et al. Antibacterial applications of graphene oxides: structure-activity relationships, molecular initiating events and biosafety [J]. Science Bulletin, **2018**, 63 (2): 133-142.

[51] Chong Yu, Ge Cuicui, Fang, et al. Light-enhanced antibacterial activity of graphene oxide, mainly via accelerated electron transfer [J]. Environmental science & technology, **2017**, 51 (17): 10154-10161.

[52] Hu Wenbing, Peng Cheng, Luo Weijie, et al. Graphene-Based Antibacterial Paper [J]. Acs Nano, **2010**, 4 (7): 4317-4323.

[53] Mangadlao Joey, M Santos C, J L Felipe M, et al. On the antibacterial mechanism of graphene oxide (GO) Langmuir-Blodgett films [J]. Chemical Communications, **2015**, 51 (14): 2886-2889.

[54] Pandit Subhendu, Karunakaran, Boda Sunil Kumar, et al. High Antibacterial Activity of Functionalized Chemically Exfoliated MoS_2 [J]. ACS applied materials & interfaces, **2016**, 8 (46): 31567-31573.

[55] Tian Tengfei, Shi Xiaoze, Cheng Liang, et al. Graphene-based nanocomposite as an effective, multifunctional, and recyclable antibacterial agent [J]. ACS applied materials & interfaces, **2014**, 6 (11): 8542-8548.

[56] Krishnamoorthy Karthikeyan, Veerapandian Murugan, Zhang Ling-He, et al. Antibacterial Efficiency of Graphene Nanosheets against Pathogenic Bacteria via Lipid Peroxidation [J]. J. Phys. Chem. C, **2012**, 116 (32): 17280-17287.

[57] Zou Fengming, Zhou Hongjian, Young Jeong Do, et al. Wrinkled Surface-Mediated Antibacterial Activity of Graphene Oxide Nanosheets [J]. Acs Applied Materials Interfaces, **2017**, 9 (2): 1343-1351.

[58] Li Ruibin, D. Mansukhani Nikhita, Guiney Linda, et al. Identification and Optimization of Carbon Radicals on Hydrated Graphene Oxide for Ubiquitous Antibacterial Coatings [J]. Acs Nano, **2016**, 10 (12): 10966.

[59] Khand M. Shahnawaz, Abdelhamida Hani Nasser, Wu Hui-Fen. Near infrared (NIR) laser mediated surface activation of graphene oxide nanoflakcs for efficient antibacterial, antifungal and wound healing treatment. Colloids and Surfaces B-Biointerfaces. **2015**. 127, 281-291.

[60] Liu Shaobin, Hu Ming, Zeng Tingying (Helen), et al. Lateral dimension-dependent antibacterial activity

of graphene oxide sheets [J]. Langmuir the Acs Journal of Surfaces Colloids, **2012**, 28 (33): 12364-12372.

[61] Karahan Hüseyin Enis, Wiraja Christian, Xu Chenjie, et al. Graphene Materials in Antimicrobial Nanomedicine: Current Status and Future Perspectives [J]. Advanced healthcare materials, **2018**, 7 (13): 1701406

[62] Liu Shaobin, Zeng Tingying, Hofmann Mario, et al. Antibacterial activity of graphite, graphite oxide, graphene oxide, and reduced graphene oxide: membrane and oxidative stress [J]. Acs Nano, **2011**, 5 (9): 6971-6980.

[63] Huang Po-Jung and Liu Juewen. Molecular beacon lighting up on graphene oxide [J]. Analytical Chemistry, **2012**, 84 (9): 4192-4198.

[64] Bart Cornelissen, Sarah Able, Veerle Kersemans, et al. Nanographene oxide-based radioimmunoconstructs for in vivo targeting and SPECT imaging of HER2-positive tumors [J]. Biomaterials, **2013**, 34 (4): 1146-1154.

[65] Cao Tianye, Zhou Xiaobao, Zheng Yingying, et al. Chelator-Free Conjugation of 99mTc and Gd^{3+} to PEGylated Nanographene Oxide for Dual-Modality SPECT/MR Imaging of Lymph Nodes [J]. ACS applied materials interfaces, **2017**, 9 (49): 42612-42621.

[66] Sheng Zonghai, Song Liang, Zheng Jiaxiang, et al. Protein-assisted fabrication of nano-reduced graphene oxide for combined in vivo photoacoustic imaging and photothermal therapy [J]. Biomaterials, **2013**, 34 (21): 5236-5243.

[67] Sun Xiaoming, Liu Zhuang, Welsher Kevin, et al. Nano-graphene oxide for cellular imaging and drug delivery [J]. Nano Research, **2008**, 1 (3): 203-212.

[68] Sheng-Jen Cheng Hsien-Yi Chiu, Priyank V. Kumar, Kuan Yu Hsieh, Jia-Wei Yang You-Rong Lin, Yu-Chih Shena, and Guan-Yu Chen. Simultaneous drug delivery and cellular imaging using graphene oxide. Biomaterials Science. **2018**. 6, 813-819.

[69] Pan Jiabao, Yang Yang, Fang Wenjuan, et al. Fluorescent Phthalocyanine-Graphene Conjugate with Enhanced NIR Absorbance for Imaging and Multi-Modality Therapy. ACS Applied Nano Materials. **2018**. 1, 2785-2795.

[70] Lu Chang, Huang Po-Jung Jimmy, Liu Biwu, et al. Comparing Graphene Oxide and Reduced Graphene Oxide for DNA Adsorption and Sensing [J]. Langmuir, **2016**, 32 (41): 10776-10783.

[71] Yuan Hongbo, Qi Junjie, Xing Chengfen, et al. Graphene-Oxide-Conjugated Polymer Hybrid Materials for Calmodulin Sensing by Using FRET Strategy [J]. Advanced Functional Materials, **2015**, 25 (28): 4412-4418.

[73] Ren Hui, Kulkarni Dhaval D., Kodiyath Rajesh, et al. Competitive adsorption of dopamine and rhodamine 6G on the surface of graphene oxide [J]. Acs Applied Materials Interfaces, **2014**, 6 (4): 2459-2470.

[74] Elkhenany Hoda, Amelse Lisa, Lafont Andersen, et al. Graphene supports in vitro proliferation and osteogenic differentiation of goat adult mesenchymal stem cells: potential for bone tissue engineering [J]. Journal of Applied Toxicology, **2015**, 35 (4): 367-374.

[75] Wei Changbo, Liu Zifeng, Jiang Fangfang, et al. Cellular behaviours of bone marrow-derived mesenchymal stem cells towards pristine graphene oxide nanosheets [J]. Cell Proliferation, **2017**, 50 (5).

[76] Kim Jangho, Seonwoo Hoon, Cho ChongSu, et al. Bioactive effects of graphene oxide cell culture substratum on structure and function of human adipose-derived stem cells [J]. Journal of Biomedical Materials Research Part A, **2013**, 101 (12): 3520-3530.

[77] Kolanthai Elayaraja, Pugazhendhi Abinaya Sindu, Khajuria Deepak Kumar, et al. Graphene Oxide-A

Tool for Preparation of Chemically Crosslinking Free Alginate-Chitosan-Collagen Scaffold for Bone Tissue Engineering [J]. Acs Appl Mater Interfaces, **2018**, 10 (15): 12441-12452.

[78] Kim Jangho, Kim Yang Rae, Kim Yeonju, et al. Graphene-incorporated chitosan substrata for adhesion and differentiation of human mesenchymal stem cells [J]. Journal of Materials Chemistry B, **2013**, 1 (7): 933-938.

[79] Sorina Dinescu, Mariana Ionita, Andreea Madalina Pandele, et al. In vitro cytocompatibility evaluation of chitosan/graphene oxide 3D scaffold composites designed for bone tissue engineering [J]. Biomed Mater Eng, **2014**, 24 (6): 2249-2256.

[80] Zhou Tengfei, Li Guo, Lin Shiyu, et al. Electrospun Poly (3-Hydroxybutyrate-Co-4-Hydroxybutyrate) /Graphene Oxide Scaffold: Enhanced Properties and Promoted in Vivo Bone Repair in Rats [J]. Acs Appl Mater Interfaces, **2017**, 9 (49): 42589-42600.

[81] La Wan-Geun, Kwon Sun-Hyun, Lee Tae-Jin, et al. The effect of the delivery carrier on the quality of bone formed via bone morphogenetic protein-2 [J]. Artificial Organs, **2012**, 36 (7): 642-647.

[82] La Wan-Geun, Park Saibom, Yoon Hee-Hun, et al. Delivery of a therapeutic protein for bone regeneration from a substrate coated with graphene oxide [J]. Small, **2013**, 9 (23): 4051-4060.

[83] Xie Youtao, Li Hongqing, Zhang Chi, et al. Graphene-reinforced calcium silicate coatings for load-bearing implants [J]. Biomedical Materials, **2014**, 9 (2): 025009.

[84] Li Ming, Liu Qian, Jia Zhaojun, et al. Graphene oxide/hydroxyapatite composite coatings fabricated by electrophoretic nanotechnology for biological applications [J]. Carbon, **2014**, 67 (34): 185-197.

第8章
CHAPTER 8

氧化石墨烯在水处理中的应用研究

GO 表面有着丰富的羟基、羧基、环氧基、羰基等亲水性的活性基团,且片层间距较大,使得 GO 具有超大比表面积和超强的离子交换能力。GO 的结构与水通蛋白相类似,而蛋白质本身具有优异的离子识别功能,由此可推断 GO 在分离、过滤及仿生离子传输等领域可能具有潜在的应用[1~3]。GO 经过超声可以稳定地分散在水中,再通过传统成膜方法如旋涂、滴涂和真空抽滤等处理后,GO 微片可呈现肉眼可见的层状薄膜堆叠,在薄膜的层与层之间形成具有选择性的二维纳米通道,可精确地分离特定尺寸范围内的目标离子和分子。除此之外,GO 由于片层间存在较强的氢键,力学性能优异,易脱离基底而独立存在。基于 GO 薄膜制备方法简单、成本低、超大比表面积和超强的离子交换能力等优点,在水净化领域具有广阔的应用空间[4~6]。

本节综述 GO 层状渗透膜对气体、水及离子的传质行为,并展望了 GO 在膜分离、海水淡化和污染物去除等环境应用中的机遇和挑战。

8.1 氧化石墨薄膜对溶液中离子的传输

基于当前严重的淡水危机,膜分离技术在海水淡化以及污水处理等领域具有能耗低、分离效率高、过程简单等不可比拟的优势,被认为是当前最有发展前景的水处理技术[7]。目前超滤膜、纳滤膜、反渗透膜等膜技术,已经成功地应用到水处理的各个领域,引起越来越多的研究关注[8~11]。GO 薄膜在海水淡化领域的主要应用是去除海水中的盐离子,因此探究 GO 薄膜的离子传质行为具有重要的意义。

从理论上讲,在 GO 的纳米孔道中,分布着氧化区域和纳米 sp^2 杂化碳区域,水分子在通过氧化区域时能够与含氧官能团形成氢键,增加了水流动阻力,而在杂化碳区域水流阻力很小。芳香碳网中形成的大多数通路被含氧官能团有效阻挡,从而分离海水中 Na^+ 和 Cl^- 等小分子物质[12,13]。相比于其他纳米材料,GO 为快速水输送提供了较多优越性能,如部分区域光滑无摩擦的表面,超薄的厚度和超高的机械强度,都有助于提高水的渗透性[14~16]。

Kim 等[17]人研究了多孔石墨烯膜在海水淡化中的应用,得到以下结论:石墨烯片层上一定尺寸的孔径可允许水分子透过,同时有效截留氯化钠 NaCl,水通量比传统的反渗透膜高出几个数量级,在海水淡化上具有很好的应用前景。但如何制备高性能石墨烯基反渗透膜还需要进一步的研究:一种可能的方法是在 GO 片层间通过共价交联方式引入小分子,以克服 GO 片层之间氢键作用力,从而减小片层孔隙之间的距离,在一定程度上可提高 Na^+ 和 K^+ 等离子的截留效率。

Sun 等[18,19]人探究了盐溶液中不同离子的渗透行为,如图 8.1、图 8.2 所示。研究人员通过滴涂法制备出微米级（5μm）的 GO 薄膜,并在 GO 膜两侧加入等量盐溶液和水,考察 GO 膜的选择渗透性和水净化特性。实验结果显示,不同金属盐的相对传导率发生了显著变化。钠盐短时间内很快通过 GO 膜,而重金属盐较慢,铜盐和有机污染物（如罗丹明 B）则不能通过,不同阴离子自由基的钠盐在近似的顺序中显示出不同的渗透率。由此可知,从铜盐和有机物中能有效分离出钠盐。进一步地,选取水合半径顺序为 $Mn^{2+}>Cd^{2+}>Cu^{2+}>Na^+$ 的盐溶液进行实验,结果表明离子渗透与其水合半径并没有直接关系（$Na^+>Mn^{2+}>Cd^{2+}>Cu^{2+}$）,而金属盐离子和 GO 片上官能团的不同相互作用则具有重要影响。研究者继续测试了相同浓度的盐溶液的渗透行为,渗透结果为 $NaOH>NaHSO_4>NaCl>NaHCO_3$,说明盐溶液中的阴离子也会对渗透结果产生影响。参照图 8.2(c) 和 (d) 所示,通过对不同盐离子的渗透机理研究可知,主族金属阳离子更易与 sp^2 基团发生作用,过渡金属阳离子更易与含氧官能团发生作用,因而导致 GO 膜具有不同的离子选择性。当 GO 膜被氨基酸目标官能团修饰时,离子的渗透性能减弱,说明可以通过在 GO 膜上修饰目标官能团来调制离子的运输速率。

GO 膜在水处理中的分离机理尚存在诸多争议。一种观点认为通过尺寸筛分以及带电的目标分离物与纳米孔之间的静电排斥机理实现分离。GO 膜的分离通道主要由两部分构成:a.GO 分离膜中不规则褶皱结构形成的半圆柱孔道；b.GO 分离膜片层之间的空隙。除此之外,由 GO 结构缺陷引起的纳米孔道对于水分子的传输提供了额外的通道[20~23]。

Mi 等[24]人研究认为,干态下通过真空过滤制备的 GO 片层间隙的距离约为 0.3nm,使得仅有单层排列的水蒸气可以渗透通过纳米通道。通过在 GO 纳米片之间夹入适当尺寸的间隔物来调节 GO 间距,可以制造广谱的 GO 膜,每个膜能够精

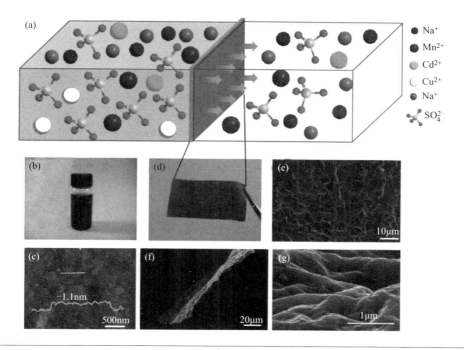

图 8.1 盐溶液中不同离子在 GO 膜中渗透行为

(a) 不同离子通过 GO 膜的渗透过程示意；(b) 用于渗透实验的 GO 分散液；(c) GO 片层的 AFM 图像；(d) GO 分散液滴涂法制备的独立 GO 膜；(e)～(g) GO 膜的表面、横截面及放大面的 SEM 图像[18]

确地分离特定尺寸范围内的目标离子和分子。水合作用力使得溶液中 GO 片层间隙的距离增大到 1.3nm，真正有效、可自由通过的孔道尺寸为 0.9nm，计算出水合半径小于 0.45nm 的物质可以通过 GO 膜片，而水合半径大于 0.45nm 的物质被截留。例如，脱盐要求 GO 的层间距小于 0.7nm，以从水中筛分水合 Na^+（水合半径为 0.36nm）。通过部分还原 GO 以减小水合官能团的尺寸或通过将堆叠的 GO 纳米片与小尺寸分子共价键合以克服水合力，可以获得这种小间距。与此相反，如果要扩大 GO 的层间距至 1~2nm，可在 GO 纳米片之间插入刚性较大的化学基团或聚合物链（例如聚电解质），从而使 GO 膜成为水净化、废水回收、制药和燃料分离等应用的理想选择。如果使用更大尺寸的纳米颗粒或纳米纤维作为插层物，可以制备出间距超过 2nm 的 GO 膜，以用于生物医学应用（例如人工肾和透析），这些应用需要大面积预分离生物分子和小废物分子。

利用化学交联和物理手段调控 GO 基膜片上的褶皱和片层间的距离是制备石墨烯基纳滤膜的主要手段。由于 GO 片层间隙距离小，Gao 等[25]人利用真空过滤法在石墨烯片层间加入 SWCNTs，GO 片层间的距离明显增加，水通量可达到

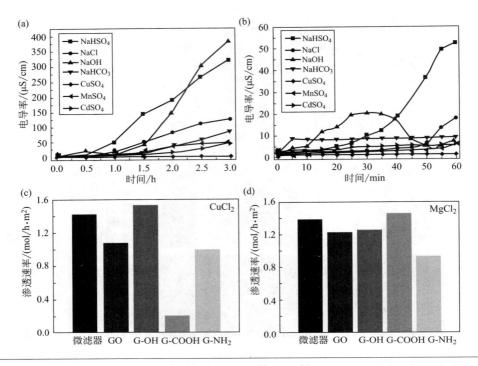

图 8.2 不同盐离子透过 GO 的变化趋势及 Mg^{2+} 和 Cu^{2+} 透过不同功能化石墨烯膜对比

(a)、(b) 不同盐离子透过 GO 薄膜随时间变化趋势[18];

(c)、(d) Mg^{2+}、Cu^{2+} 透过 GO 和不同功能化的石墨烯薄膜对比[19]

6600~7200L/(m^2·h·MPa),大约是传统纳滤膜水通量的 100 倍,对于染料的截留率达到 97.4%~98.7%。Joshi 等[26]人研究了真空抽滤 GO 分散液制备微米级厚度层状 GO 薄膜的渗透作用。一系列实验表明,GO 膜在干燥状态下是真空压实的,但作为分子筛浸入水中后,能够阻挡所有水合半径大于 0.45nm 的离子。半径小于 0.45nm 的离子渗透速率比自由扩散高出数千倍,这种行为是由纳米毛细管网络引起的。异常快速渗透归因于作用在石墨烯毛细管内部离子的毛细管样高压。GO 薄膜的这一特性在膜分离领域具有重要的应用价值。

GO 基纳滤膜水通量远远大于传统的纳滤膜,但是 GO 纳滤膜对盐离子的截留率还有待提高。Han 等[27]人利用过滤法在 GO 片层中间混合加入多壁碳纳米管,复合膜的通量达到 113L/(m^2·h·MPa),提高了对于盐离子的截留率,例如对于 Na_2SO_4 截留率可达到 83.5%。Sun 等[28]人提出了一种精确可控的基于 GO 的复合渗透膜的设计思路,通过将单层二氧化钛(TO)纳米片嵌入具有温和紫外(UV)光照还原的 GO 层压材料中,所制备的 RGO/TO 杂化膜表现出优异的水脱盐性能。另外,TO 具有光致亲水特性,可保证高的水流速率,在没有外部流体静压的情况下，与 GO/TO 情况相比，通过 RGO/TO 杂化膜的离子渗透率可降低至 0.5%，而使用同位素标记技术测量的水渗透率可保持为原来的 60%。RGO/TO 杂化膜优异的脱盐性能，表明 TO 对 GO 的光致还原作用有助于离子的有效排斥，而在紫外光照射下光诱导 TO 的亲水转化是保留优异的水渗透性的主要原因。这种复合薄膜制备方法简单，在水净化领域具有潜在应用。Su 等[29] 人利用氢碘酸和抗坏血酸对 PET 基底上的多层 GO 薄膜进行化学还原，得到 30nm 厚的 RGO 薄膜，并测试了其渗透性能。实验发现，对 He 原子和水分子完全不能透过。而厚度超过 100nm 的 RGO 薄膜对几乎所有气体、液体和腐蚀性化学试剂（如 HF）是高度不可渗透的。特殊的阻隔性能归因于石墨烯层压结构的高度石墨化和在还原过程中几乎没有结构损坏。与此结果相反，Liu 等[30] 人证明了通过 HI 蒸汽和水辅助分层，是制备独立式超薄 RGO 膜的简便且可重复的方法，利用 RGO 膜的毛细管力和疏水性，可通过水实现最终的分层。研究人员采用真空抽滤在微孔滤膜基底上制备了厚度低至 20nm 的独立式 RGO 薄膜，进而对 NaCl 溶液进行测试。发现这种 RGO 膜不仅具有高水通透速率，且和商业膜相比 GO 膜具有更高的脱盐性能。出现上述截然不同的结果的可能原因是，GO 可能是只有部分被还原，还有可能是在操作过程中破坏了 GO 的微观结构，具体原因尚需进一步研究确认。

8.2 氧化石墨薄膜对气体和水分子的传输

2012 年，Nair 等[31] 人通过旋涂法制备了能够有效阻挡气体（包括氦气、氢气、氮气等）和有机溶剂蒸汽的微米级 GO 薄膜。然而，这样的薄膜材料却使水分子能够无阻碍渗透过膜，表明 GO 膜具有潜在的选择性分离作用。另外，Abraham 等[14] 人又通过研究表明，经不同湿度处理后，GO 膜的层间距在 0.64~0.98nm 范围内可控。通过将这样厚度为 100μm 的 GO 膜进行环氧树脂封装实现物理限制，能够有效抑制 GO 膜在水中的溶胀，从而精确控制离子筛分，实现对 NaCl 高达 97% 的截留率。更小的通道尺寸使离子渗透速率以指数形式降低，但水分子的传质速度并没有较大影响，这归因于水分子在石墨烯毛细通道中较低的能量势垒。值得注意的是，该实验中由于物理封装 GO 膜的条件限制，实验测试过程并不是传统意义上的"平板膜"渗透方式，而是以 100μm 厚的膜截面为截留平面所进行的"截面二维通道"渗透方式。Boukhvalov 等[32] 基于第一性原理密度泛函理论，建立了水和 GO 混合系统的原子模型，揭示了堆积在 GO 内部水的异常渗透行为，发现在 GO 层片之间形成的六方冰双层及其在薄片边缘的冰融化过渡对于实现整个堆垛结构中完美的水渗透作用至关重要；进而通过对 GO 化学还原处理或施加一定的压力来调节 GO 层片间距，可对水流速进行有效控制。通过不同的堆垛方法来控制 GO 片层间的孔道，对于厚度为 3~10nm 的 GO 微片进行组装，可实现对气体的高选择性。在较高的相对湿度下，制备较好的 GO 薄膜可以实现 CO_2/N_2 两种气体的高效分离[33]。Li 等[34]人利用真空抽滤法制备了厚度约 1.8nm、具有一定的结构缺陷的超薄 GO 膜，对 H_2/CO_2 及 H_2/N_2 混合气体可分别实现高达 3400 和 900 的选择性，比最先进的微孔滤膜高出 1～2 个数量级。实验表明，H_2/CO_2 的分离选择性随着温度的升高而降低，但即使在 100℃温度下，18nm 厚的膜 H_2/CO_2 选择性仍为 250。作者认为，由于 CO_2 分子在结构缺陷中的紧密配合，使 CO_2 通过 GO 膜的渗透性增加比 H_2 更快。利用超薄 GO 膜从混合气体中分离出 H_2 可能会有一定的应用前景。

以上研究结果表明，GO 薄膜对气体通过具有选择性，这看似同 Nair 等人[31]的结果矛盾，其实不然。由于前者制备的 GO 薄膜皆为数纳米厚、较薄，连续的纳米毛细管网络尚未形成，气体可通过缺陷进行渗透；而后者的 GO 薄膜为微米厚度，具有连续的 GO 毛细管通道网络，缺陷被互相覆盖，所以气体无法透过[35～37]。

以上研究都是针对气体和水蒸气的渗透性能研究，关于液态水的渗透现象要比气体复杂得多。这是因为被 GO 薄膜所分离的源液和滤液之间，在宏观和微观尺度上都没有可分辨的不同点。Sun 等[38]人基于同位素示踪法解决了这一问题，在无外部静水压力情况下研究了液态水在微孔滤膜基底上 GO 薄膜的透过行为。他们利用重水标记液态水，根据其渗透特性来研究液态水的渗透特性。在原液中加入 $MgCl_2$，并用 D_2O 示踪剂标记 0.1M $MgCl_2$ 原液和滤液以研究离子存在下的正向水输运，结果表明当离子溶解在原溶液中时，水渗透速率稍微降低。作者认为，纳米毛细管中存在的离子可能会导致有序的冰双层发生一些变形，进而产生水渗透速率的减小。接着，用质量分数为 30% 的 D_2O 标记排水溶液以研究在离子存在下从排水到水源的水渗透。结果显示，当原液中存在 0.1M $MgCl_2$ 时，从漏极到源极的水渗透比从源极到漏极的渗透稍快，液态水可以通过基于石墨烯的纳米通道提供超快速渗透，离子的渗透系数和水的渗透系数在同一个数量级。渗透系数比孔径小于微米级的通道高 4～5 个数量级，且离子主要由水流运输，离子和纳米毛细管壁之间的微妙相互作用也在加速离子迁移中起作用。这一现象表明 GO 薄膜具有半渗透性，同时水分子在通过 GO 薄膜时倾向于逆离子梯度方向传输。

最近，Zhou 等[39]人开发出一种石墨烯膜，可以精确控制水流，从超快渗透到完全阻塞。这样石墨烯能够在处理液体和气体时形成可调谐的滤光器，甚至形成完美的屏障。该器件通过 GO 膜的电控制水渗透，在电绝缘 GO 膜中嵌入导电丝，通过这些纳米丝的电流产生一个大的电场，电离水分子，从而控制通过膜中的石墨烯毛细管的水传输。具体做法是：在多孔银（Ag）基板上制备 GO 膜，表面沉积薄（约 10nm）金（Au）膜制备 Ag/GO/Au 夹层膜结构。将夹层结构胶合到具有圆孔（直径约 10mm）的塑料圆盘上，然后用于密封充水钢容器以使装置暴露于水蒸气并使用重量分析仪测量水渗透。结果表明，在石墨烯毛细管中观察到的水流量随离子浓度的增加而减少，这可能是由于水合作用；在石墨烯毛细管中，随着离子数量越来越多，水往往停留在毛细管中，从而减少了水的流动。还值得一提的是，在分子动力学模拟中，用于降低水流量的离子浓度被高估。GO 的层间通道中可用的自由空间<1nm，GO 毛细管内的官能团可吸收离子并使通道更亲水。然而，分子动力学模拟表明，GO 膜层间通道内的离解水分子可以显著影响水通过膜的渗透。此技术不仅仅局限于控制水的流动，相同的膜还可以用作吸附剂或海绵。如果施加电流，即使在沙漠条件下，吸附在膜上的水也可以保留在膜中。

8.3 氧化石墨烯对重金属离子的吸附

当前社会的快速发展造成了严重的重金属离子污染，重金属离子毒性大、分布广、难降解，一旦进入生态环境，严重威胁人类的生命健康。目前，含重金属离子废水的处理方法主要有化学沉淀法、膜分离法、离子交换法、吸附法等。而使用纳米材料吸附重金属离子成为当前科研人员的研究热点。相对活性炭、碳纳米管等碳基吸附材料，GO 的比表面积更大，表面官能团（如羧基、环氧基、羟基等）更为丰富，具有很好的亲水性，可以与金属离子作用富集分离水相中的金属离子；同时，GO 片层可交联极性小分子或聚合物制备出 GO 纳米复合材料，吸附特性更加优异[40~44]。

GO 吸附金属离子的作用机制主要包括：GO 与金属离子的静电作用、GO 表面官能团与金属离子发生络合反应、离子交换及配体交换等。

8.3.1 静电作用

GO 表面含有-OH 和-COOH 等丰富的官能团，在水中可发生去质子化等反应带有负电荷，由静电作用将金属阳离子吸附至表面；相反的，如果水中 pH 等环境因素发生变化，GO 表面也可携带正电荷，则与金属离子产生静电斥力，二者之间的吸附作用大大减弱。而静电作用的强弱与 GO 表面官能团产生的负电荷相关，受环境 pH 值的影响较明显。Wang 等[45]人的研究证明，在 $pH > pH_{pzc}$ 时（$pH_{pzc}=3.8$），GO 表面的官能团可发生去质子化反应而带负电，可有效吸附铀离子 U(Ⅵ)，吸附量可达到 1330mg/g。

8.3.2 络合作用

GO 表面的-OH 和-COOH 等官能团含有孤对电子，可作为配位体与具有空的价电子轨道的金属离子发生络合反应，生成不溶于水的络合物，从而有效去除溶液中的金属离子。Madadrang 等[46]人制得乙二胺四乙酸（EDTA）/GO 复合材料（记作 EDTA-GO），通过研究发现其对金属离子的吸附机制主要为络合反应，即

GO 的表面官能团与水中的金属离子反应形成复杂的络合物，具体过程如图 8.3 所示。由于形成的络合物不溶于水，可通过沉淀等作用分离去除水中的金属离子。

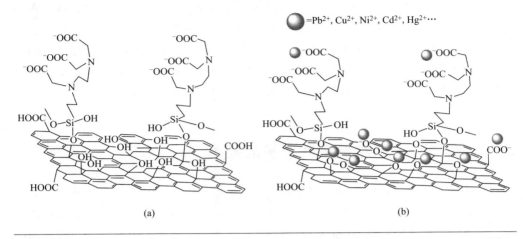

图 8.3 EDTA-GO 的化学结构图（a）和 EDAT-GO 与各重金属离子间的络合作用（b）[46]

8.3.3 离子交换作用

GO 的表面官能团可与溶液中的金属离子进行交换反应，从而将金属离子固定在 GO 片层结构上。Yang 等[47]人研究了 GO 对 Sr^{2+} 的吸附作用，实验表明其主要吸附机制为离子交换，在 pH 值较低的环境中，GO 表面的含氧官能团被去质子化，与吸附在 GO 上面的 Na^+/H^+ 与 Sr^{2+} 发生离子交换反应，即①$2\equiv S-OH + Sr^{2+} \rightarrow (S-O)_2-Sr + 2H^+$；②$2\equiv S-ONa + Sr^{2+} \rightarrow (\equiv S-O)_2-Sr + 2Na^+$。

8.3.4 配体交换作用

配体交换作用即：GO 上原有的配位体被溶液中的金属离子所取代，并以配位键的形式生成不溶于水的配合物，最终通过简单的过滤即可从溶液中去除。Tang 等[48]人对 Fe 与 GO（质量比为 1∶7.5）复合及 Fe 与 Mn（摩尔比为 3∶1）复合的 GO/铁-锰复合材料（GO/Fe-Mn）进行了吸附研究，通过一系列的实验表明，GO 对 Hg^{2+} 的吸附机理主要是配体交换作用，其最大吸附量达到 32.9mg/g。Hg^{2+} 可在水环境中形成 $Hg(OH)_2$，与铁锰氧化物中的活性位点（如-OH）发生配体交换作用，从而将 $Hg(OH)_2$ 固定在 GO/铁-锰复合材料上，达到去除水环境中 Hg^{2+} 的目的。

GO 经一定功能化处理后可发挥更大的性能优势，例如大比表面积、高敏感度和高选择性等，这些特性对于 GO 作为吸附剂吸附水环境中的金属离子有着重要的作用。

Zhang 等[49,50]人采用乙二胺（EDA）和超支化聚乙烯亚胺对 GO 进行修饰，制备出具有较高正电荷和较低运输阻力的纳米骨架复合膜，纯水通过此膜的水通量可达 5.01L/(m²·h·bar)。在 1bar（1bar=0.1MPa）压力下进行死端过滤，经过滤 2h 后，Cd^{2+}、Zn^{2+}、Ni^{2+} 和 Pb^{2+} 的去除率达到 90.5%、97.4%、96.0% 和 95.7%，而其初始重金属离子浓度是 1000mg/L，可见 GO 经修饰后吸附能力有很大的提升。Thong 等[51]人经研究又成功地采用逐层组装法将 GO 沉积在中空纤维膜上，使此复合膜的纯水通量达 4.7L/(m²·h·bar)，Pb^{2+}、Ni^{2+}、Zn^{2+} 的去除率均高于 95%。

Mukherjee 等[52]人将 GO 混合到聚砜（Polysulfone，简称 PSF）中制备出复合基质膜，考察了跨膜压差、错流速率以及溶液 pH 值对重金属去除率的影响规律，并探究出了最佳实验条件。当初始重金属离子浓度为 50mg/L 时，TMP 为 414kPa，Pb^{2+}、Cu^{2+}、Cd^{2+}、Cr^{6+} 的去除率在 90%～96%。

Reza 等[53]人通过湿式相转移法制备了 PSF/GO 复合膜，考察了砷酸根离子（AsO_2^{2-}）的去除率和水通量。其初始浓度为 300μg/L，TMP 为 4bar，pH 为 8.5。当 GO 添加量为 1% 时，AsO_2^{2-} 的去除率为 82.3%，水通量为 43.1L/(m²·h)，去除机理主要是电荷之间的 Donnan 排斥。

Liu 等[54]人通过原位液相沉积的方法将 $\delta\text{-}MnO_2$ 纳米薄片与石墨烯纳米片进行自组装，制备出 3D 石墨烯/$\delta\text{-}MnO_2$ 气凝胶，并对吸附动力学和对重金属离子吸附性能进行了深入研究。结果显示，复合气凝胶对 Pb^{2+}、Cd^{2+}、Cu^{2+} 的吸附量分别可达 643.26mg/g、250.31mg/g 和 228.46mg/g，且用 HCl 和 KOH 作为解吸剂处理后可以再生，可循环利用 8 次，不会产生二次污染。

8.4 氧化石墨烯对水中有机污染物的吸附

工业化和城市化导致天然地表水体中的有毒化学品排放，其中包括酚类、油污、抗生素、农药和腐植酸等有机物，这些污染物在制药，石化，染料，农药等行业的废水中广泛检测到。许多研究集中在从水溶液中有效去除这些有毒污染物，如光催化，吸附和电解[55~58]。在这些方法中，由于吸附技术低成本、高效率和易于操作，优于其他技术。

与传统的膜材料不同，GO 作为碳质材料与有机分子的相互作用机理差异很大。新的界面作用可在 GO 膜内引入独特的传输机制，导致更有效地从水中去除有机污染物。石墨烯和 GO 对有机物的吸附机理的研究表明，疏水作用、π-π 键交互作用、氢键、共价键和静电相互作用会影响石墨烯和 GO 对有机物的吸附能力[59~61]。

Oh 等[62]人报道了多层 GO 膜在不同 pH 水平下去除水中有机物质的系统性能评价和机理研究。该研究采用逐层组装法制备了 PAH/GO 双层膜，对典型单价离子（Na^+，Cl^-）和多价离子（SO_4^{2-}，Mg^{2+}）以及有机染料（亚甲基蓝 MB，罗丹明 R-WT）和药物和个人护理品［如三氯生（TCS），三氯二苯脲（TCC）］在反渗透膜系统中通过 GO 膜的行为进行研究。结果发现，在 pH＝7 时，无论其电荷、尺寸或疏水性质如何，GO 膜能够高效去除多价阳离子/阴离子和有机物，但对于单价离子的去除率较低。传统的纳滤膜通常带负电，且只能去除带有负电荷的多价离子和有机物。随着 pH 的变化，GO 膜的关键性质（例如电荷，层间距）发生显著变化，导致不同的 pH 依赖性界面现象和分离机制，一些有机物（例如三氯二苯脲）的分子形状由于这种有机物与 GO 膜的碳表面的迁移性和 π-π 相互作用而极大地影响了它们的去除。

该研究表明，当 pH 由 7 增加至 10，MB 的去除率从 97％ 急剧下降到 62％，这是由于正氨基团的中和使 GO 膜在 pH＝10 下带负电，从而对带正电的 MB 不具备排斥能力。此外，PAH 的正电荷损失使其在负电 GO 层间发生电荷排斥，导致 GO/PAH 双层溶胀，层间距增大，进而降低了 pH 值为 10 时的 MB 的去除率。但是，带负电的 R-WT 在 pH 为 7 和 10 时去除率接近 100％，在 pH＝4 时下降至 70％。推测可能原因是，在 pH＝4 时，由于羧基的质子化，GO 膜损失了大部分负电荷，削弱了对 R-WT 的电荷排斥，从而降低了去除率。此外，由于 GO 膜内的大量正电荷引起的 GO/PAH 双层溶胀增大了其层间距，减小了尺寸排阻，从而进一步降低了 pH＝4 下 R-WT 的去除。

在 pH＝10 的 20mM NaCl 中，TCS 的去除率仅为 37％。因为中性化合物的电离会降低其疏水性，从而降低了其与 GO 膜的亲和力，促进了 TCS 通过 GO 膜的更快迁移。相反，在 pH 为 4 和 7 时，TCS 在 GO 膜中具有高度疏水性和较低的流动性，导致更高的去除率。TCC 的去除几乎不依赖于电荷效应，主要归因于其疏水性和分子形状[63]。TCC 的 pKa 为 12.7，因此在 pH 为 4、7 和 10 时保持中性，表明电荷效应在其与 GO 膜的相互作用中不起作用。与其他受 pH 影响的有机物不同，TCC 在所有 pH 水平下的去除率始终接近 100％。TCC 是唯一在同一平面上具有两个苯环的有机分子，二维结构促进 π-π 键与 GO 表面的未氧化区域相互作用，从而明显降低 TCC 在 GO/PAH 双层膜之间的迁移率[64]。因此，尽管 TCS 和 TCC 具有相似的疏水性，GO 膜对 TCC 的去除率要高得多，且不依赖 pH 值。当处于中性形式时，离子强度对 TCS 或 TCC 的去除几乎没有影响。然而，在 pH＝10 时，较高的离子强度（例如 100mM NaCl）会增加 TCS 的排斥，可能是因为加入的盐导致表面电荷的屏蔽效应减弱了静电排斥[65,66]，使疏水作用增加了对 TCS 的排斥。

Pei 等[67]人分析了石墨烯和 GO 吸附 1，2，4-三氯苯（TCB）、2，4，6-三氯苯酚（TCP）、2-萘酚和萘（NAPH）的吸附机理。结果表明，在碱性 pH 条件下，

2-萘酚对 GO 具有较高的吸附能力。理论计算认为这是因为阴离子 2-萘酚的 p 电子密度高于中性 2-萘酚，促进了与石墨烯形成的 π-π 键相互作用。对于 GO，四种芳香烃的吸附亲和力依次增加：NAPH＜TCB＜TCP＜2-萘酚。四种芳烃的吸附等温线都是非线性的，在吸附过程中除了疏水相互作用，还存在某些特定的相互作用。例如，TCP、TCB 和 2-萘酚时主要通过 π-π 键的相互作用吸附在石墨烯上，相反，TCP 和 2-萘酚对 GO 的高吸附归因于 TCP 的羟基与 2-萘酚和 GO 上的含氧官能团之间的 H 键的形成。Xu 等[68]人研究了不同条件下（如温度、pH 等）石墨烯吸附对水溶液中双酚 A（Bisphenol A，简称 BPA）的去除效果。通过伪二级动力学模型和 Langmuir 等温曲线可知 BPA 的最大吸附量在 302.15K 时为 182mg/g，与文献中的其他碳质吸附剂相比，石墨烯对 BPA 的吸附效果最佳。石墨烯优异的吸附能力主要是由于其独特的 sp^2 杂化单原子层结构，在吸附过程中可通过 π-π 相互作用和氢键使 BPA 吸附在石墨烯上。另外，热力学研究表明，吸附反应是自发和放热过程。且溶液中 NaCl 的存在可以促进吸附过程，而溶液的碱性 pH 范围和较高的温度则对吸附过程不利。

Maliyekkal 等[69]人报道了 GO 和 RGO 对水体中农药如毒死蜱、硫单和马拉硫磷的吸附，发现 RGO 的吸附性能高于 GO，吸附量分别高达 1200mg/g、1100mg/g 和 800mg/g。研究还发现，吸附过程对 pH 和溶液中的离子浓度并不敏感，吸附剂是可重复利用的，并可通过合适方法改性。根据第一性原理密度泛函理论分析表明，石墨烯-水-农药在相互作用的过程中，水起到媒介的作用，石墨烯和农药的接触相当微弱甚至不接触。

Yu 等[70]人通过将少量 Fe_3O_4 纳米颗粒掺入 GO 中制备多孔、低密度（4.6mg/cm^3）的磁性 GO 海绵（MGOS）用于四环素吸附，在吸附抗生素四环素中显示出超高的吸附能力，吸附容量为 473mg/g，与 GO 相比增加了 50%。吸附速度快，可用伪二级模型描述。热力学研究表明，吸附是吸热的，驱动力是熵增加，pH 值对吸附有轻微影响，而离子强度几乎没有影响。Fe_3O_4 的小尺寸和负表面电荷，石墨烯片的高氧化度和冻干均有助于 MGOS 的高吸附容量。

8.5 小结

作为石墨烯的一种衍生物，GO 表面丰富的含氧基团使其亲水性较好，表面的负电荷较高，因此，在吸附水中有机污染物、重金属等水处理领域具有广阔的发展前景。但是目前，将石墨烯复合材料广泛应用于水处理仍存在很大的难度。首先，当前的研究主要基集中在模型污染物的基础工作，而面向实际应用的研究却很鲜见。其次，石墨烯对污染物吸附的选择性较差。另外，如何大批量、低成本制备石墨烯吸附材料并实现高效回收仍是石墨烯吸附材料实际应用的关键瓶颈。

参考文献

[1] Yazid Muhammad Noor Afiq Witri Muhammad, Sidik Nor Azwadi Che, Yahya Wira Jazair. Heat and mass transfer characteristics of carbon nanotube nanofluids: a review [J]. Renewable Sustainable Energy Reviews, **2017**, 80: 914-941.

[2] Cao Ning, Lyu Qian, Li Jin, et al. Facile synthesis of fluorinated polydopamine/chitosan/reduced graphene oxide composite aerogel for efficient oil/water separation [J]. Chemical Engineering Journal, **2017**, 326: 17-28.

[3] Chu Kyoung Hoon, Al-Hamadani Yasir, Park Chang Min, et al. Ultrasonic treatment of endocrine disrupting compounds, pharmaceuticals, and personal care products in water: A review [J]. Chemical Engineering Journal, **2017**, 327: 629-647.

[4] Nidheesh Puthiya Veetil. Graphene-based materials supported advanced oxidation processes for water and wastewater treatment: a review [J]. Environ Sci Pollut Res Int, **2017**, 24 (35): 27047-27069.

[5] Xu Weiwei, Fang Chao, Zhou Fanglei, et al. Self-Assembly: A Facile Way of Forming Ultrathin, High-Performance Graphene Oxide Membranes for Water Purification [J]. Nano Letters, **2017**, 17 (5): 2928-2933.

[6] Gupta Vinod Kumar, Agarwal Shilpi, Asif Mohammad, et al. Application of response surface methodology to optimize the adsorption performance of a magnetic graphene oxide nanocomposite adsorbent for removal of methadone from the environment [J]. Journal of colloid interface science, **2017**, 497: 193-200.

[7] Chang Yanhong, Shen Yudi, Kong Debin, et al. Fabrication of the reduced preoxidized graphene-based nanofiltration membranes with tunable porosity and good performance [J]. RSC Advances, **2017**, 7 (5): 2544-2549.

[8] Williams Christopher, Carbone Paola. Selective Removal of Technetium from Water Using Graphene Oxide Membranes [J]. Environmental Science & Technology, **2016**, 50 (7): 3875.

[9] Akbari Abozar, Sheath Phillip, Martin Samuel T., et al. Large-area graphene-based nanofiltration membranes by shear alignment of discotic nematic liquid crystals of graphene oxide [J]. Nature Communications, **2016**, 7: 10891.

[10] Yang Q, Su Yimeng, Chi Chenglong, et al. Ultrathin graphene-based membrane with precise molecular sieving and ultrafast solvent permeation [J]. Nature Materials, **2017**, 16 (12): 1198.

[11] Zhao Yali, Zhang Zhiguang, Dai Lei, et al. Enhanced both water flux and salt rejection of reverse osmosis membrane through combining isophthaloyl dichloride with biphenyl tetraacyl chloride as organic phase monomer for seawater desalination [J]. Journal of Membrane Science, **2017**, 522: 175-182.

[12] Cho Young Hoon, Kim Hyo Won, Lee Hee Dae, et al. Water and ion sorption, diffusion, and transport in graphene oxide membranes revisited [J]. Journal of Membrane Science, **2017**, 544: 425-435.

[13] Zheng Sunxiang, Tu Qingsong, J. Urban Jeffrey, et al. Swelling of Graphene Oxide Membranes in Aqueous Solution: Characterization of Interlayer Spacing and Insight into Water Transport Mechanisms [J]. Acs Nano, **2017**, 11 (6): 6440.

[14] Abraham Jijo, K S Vasu, Williams Christopher, et al. Tunable sieving of ions using graphene oxide membranes [J]. Nature Nanotechnology, **2017**, 12 (6): 546.

[15] Rollings Ryan C., T. Kuan Aaron, and A. Golovchenko Jene. Ion selectivity of graphene nanopores [J]. Nature Communications, **2016**, 7: 11408.

[16] Werber Jay R., Osuji Chinedum O., and Elimelech Menachem Materials for next-generation desalination

and water purification membranes [J]. Nature Reviews Materials, 2016, 1 (5): 16018.

[17] Kim Seungju, Ou Ranwen, Hu Yaoxin, et al. Non-swelling graphene oxide-polymer nanocomposite membrane for reverse osmosis desalination [J]. Journal of Membrane Science, 2018, 562: 47-55.

[18] Sun Pengzhan, Zhu Miao, Wang Kunlin, et al. Selective ion penetration of graphene oxide membranes [J]. Acs Nano, 2013, 7 (1): 428-37.

[19] Sun Pengzhan, Liu He, Wang Kunlin, et al. Selective Ion Transport through Functionalized Graphene Membranes Based on Delicate Ion-Graphene Interactions [J]. Journal of Physical Chemistry C, 2014, 118 (33): 19396-19401.

[20] Feng Jiandong, Liu Ke, Graf Michael, et al. Observation of ionic Coulomb blockade in nanopores [J]. Nature Materials, 2016, 15 (8): 850-855.

[21] Boya Radha, Esfandiar Ali, Wang Feng-Chao, et al. Molecular transport through capillaries made with atomic-scale precision [J]. Nature, 2016, 538 (7624): 222-225.

[22] Choi Wook, Chun Kyoung Yong, Kim Jongwoon, et al. Ion transport through thermally reduced and mechanically stretched graphene oxide membrane [J]. Carbon, 2017, 114: 377-382.

[23] Aher Ashish, Cai Yuguang, Majumder Mainak, et al. Synthesis of graphene oxide membranes and their behavior in water and isopropanol [J]. Carbon, 2017, 116: 145-153.

[24] Mi Baoxia. Materials science. Graphene oxide membranes for ionic and molecular sieving [J]. Science, 2014, 343 (6172): 740-742.

[25] Gao Shou Jian, Qin Haili, Liu Pingping, et al. SWCNT-intercalated GO ultrathin films for ultrafast separation of molecules [J]. Journal of Materials Chemistry A, 2015, 3 (12): 6649-6654.

[26] Joshi Rakesh, Carbone Paola, Wang Feng-Chao, et al. Precise and ultrafast molecular sieving through graphene oxide membranes [J]. Science, 2014, 343 (6172): 752-754.

[27] Han Yi, Jiang Yanqiu, Gao Chao. High-Flux Graphene Oxide Nanofiltration Membrane Intercalated by Carbon Nanotubes [J]. Acs Appl Mater Interfaces, 2015, 7 (15): 8147.

[28] Sun Pengzhan, Chen Qiao, Li Xinda, et al. Highly efficient quasi-static water desalination using monolayer graphene oxide | [sol] | titania hybrid laminates [J]. Npg Asia Materials, 2015, 7 (2): e162.

[29] Su Yimeng, Kravets Vasyl, Wong Swee Liang, et al. Impermeable barrier films and protective coatings based on reduced graphene oxide [J]. Nature Communications, 2014, 5: 4843.

[30] Liu Huiyuan, Wang Huanting, and Zhang Xiwang. Facile fabrication of freestanding ultrathin reduced graphene oxide membranes for water purification [J]. Advanced Materials, 2015, 27 (2): 249-254.

[31] Nair R. R., Wu H. A., Jayaram P. N., et al. Unimpeded permeation of water through helium-leak-tight graphene-based membranes [J]. Science, 2012, 335 (6067): 442-444.

[32] Boukhvalov D, Katsnelson Mikhail, Woo Son Young. Origin of anomalous water permeation through graphene oxide membrane [J]. Nano Letters, 2013, 13 (8): 3930-3935.

[33] Kim Hyo Won, Hee Wook Yoon, Seon-Mi Yoon, et al. Selective gas transport through few-layered graphene and graphene oxide membranes [J]. Science, 2013, 342 (6154): 91-95.

[34] Li Hang, Song Zhuonan, Zhang Xiaojie, et al. Ultrathin, molecular-sieving graphene oxide membranes for selective hydrogen separation [J]. Science, 2013, 342 (6154): 95-98.

[35] Heo Jiwoong, Choi Moonhyun, Chang Jungyun, et al. Highly Permeable Graphene Oxide/Polyelectrolytes Hybrid Thin Films for Enhanced CO_2/N_2 Separation Performance [J]. Scientific Reports, 2017, 7 (1): 456.

[36] Huang Kang, Liu Gongping, Jin Wanqin. Vapor transport in graphene oxide laminates and their application in pervaporation [J]. Current opinion in chemical engineering, 2017, 16: 56-64.

[37] Ying Yulong, Ying Wen, Li Qiaochu, et al. Recent advances of nanomaterial-based membrane for water purification [J]. Applied Materials Today, **2017**, 7: 144-158.

[38] Sun Pengzhan, Liu He, Wang Kunlin, et al. Ultrafast liquid water transport through graphene-based nanochannels measured by isotope labelling [J]. Chemical Communications, **2015**, 51 (15): 3251-3254.

[39] Zhou Kai-Ge, K S Vasu, T. Cherian C, et al. Electrically controlled water permeation through graphene oxide membranes [J]. Nature, **2018**, 559 (7713): 236.

[40] Peng Weijun, Li Hongqiang, Liu Yanyan, et al. A review on heavy metal ions adsorption from water by graphene oxide and its composites [J]. Journal of Molecular Liquids, **2017**, 230: 496-504.

[41] Uddin Mohammad Kashif. A review on the adsorption of heavy metals by clay minerals, with special focus on the past decade [J]. Chemical Engineering Journal, **2017**, 308: 438-462.

[42] Carolin Femina, Ponnusamy Senthil Kumar, Saravanan A, et al. Efficient Techniques for the Removal of Toxic Heavy Metals from Aquatic Environment: A Review [J]. Journal of Environmental Chemical Engineering, **2017**, 5 (3): 2782-2799.

[43] Lu Feng, Astruc Didier. Nanomaterials for removal of toxic elements from water [J]. Coordination Chemistry Reviews, **2018**, 356: 147-164.

[44] Kołodyńska Dorota, Krukowska J, Thomas P. Comparison of Sorption and Desorption Studies of Heavy Metal Ions From Biochar and Commercial Active Carbon [J]. Chemical Engineering Journal, **2017**, 307: 353-363.

[45] Wang Xiangxue, Fan Qiaohui, Yu Shujun, et al. High sorption of U (VI) on graphene oxides studied by batch experimental and theoretical calculations [J]. Chemical Engineering Journal, **2016**, 287: 448-455.

[46] J Madadrang Clemonne, Yun Kim Hyun, Gao Guihua, et al. Adsorption behavior of EDTA-graphene oxide for Pb (II) removal [J]. Acs Appl Mater Interfaces, **2012**, 4 (3): 1186-1193.

[47] Yang Shubin, Wang Xiangxue, Dai Songyuan, et al. Investigation of 90 Sr (II) sorption onto graphene oxides studied by macroscopic experiments and theoretical calculations [J]. Journal of Radioanalytical Nuclear Chemistry, **2016**, 308 (2): 721-732.

[48] Tang Jingchun, Huang Yao, Gong Yanyan, et al. Preparation of a novel graphene oxide/Fe-Mn composite and its application for aqueous Hg (II) removal [J]. Journal of hazardous materials, **2016**, 316: 151-158.

[49] Zhang Yu, Zhang Sui, and Chung Tai-Shung. Nanometric Graphene Oxide Framework Membranes with Enhanced Heavy Metal Removal via Nanofiltration [J]. Environmental Science Technology, **2015**, 49 (16): 10235-10242.

[50] Zhang Yu, Zhang Sui, Gao Jie, et al. Layer-by-layer construction of graphene oxide (GO) framework composite membranes for highly efficient heavy metal removal [J]. Journal of Membrane Science, **2016**, 515: 230-237.

[51] Thong Zhiwei, Gang Han, Cui Yue, et al. Novel nanofiltration membranes consisting of a sulfonated pentablock copolymer rejection layer for heavy metal removal [J]. Environmental Science Technology, **2014**, 48 (23): 13880-13887.

[52] Mukherjee Raka, Bhunia Prasenjit, De Sirshendu. Impact of graphene oxide on removal of heavy metals using mixed matrix membrane [J]. Chemical Engineering Journal, **2016**, 292: 284-297.

[53] Reza Rezaee, Simin Nasseri, Amir Hossein Mahvi, et al. Fabrication and characterization of a polysulfone-graphene oxide nanocomposite membrane for arsenate rejection from water [J]. Journal of Environmental Health Science Engineering, 13, 1, **2015**, 13 (1): 61.

[54] Liu Juntao, Xiao Ge, Ye Xinxin, et al. 3D graphene/δ-MnO_2 aerogels for highly efficient and reversible

removal of heavy metal ions [J]. Journal of Materials Chemistry A, **2016**, 4 (5): 1970-1979.

[55] Rastogi Monisha, Vaish Rahul. Visible light induced water detoxification through Portland cement composites reinforced with photocatalytic filler: A leap away from TiO_2 [J]. Construction Building Materials, **2016**, 120 (1 september 2016): 364-372.

[56] Umukoro Eseoghene H., Peleyeju Moses G., Ngila Jane C., et al. Photocatalytic degradation of acid blue 74 in water using $Ag-Ag_2O-ZnO$ nanostuctures anchored on graphene oxide [J]. Solid State Sciences, **2016**, 51: 66-73.

[57] Nguyen Linh Viet, Busquets Rosa, Ray Santanu, et al. Graphene oxide-based degradation of metaldehyde: Effective oxidation through a modified Fenton's process [J]. Chemical Engineering Journal, **2017**, 307: 159-167.

[58] Rostamian Rahele, Behnejad Hassan. A comparative adsorption study of sulfamethoxazole onto graphene and graphene oxide nanosheets through equilibrium, kinetic and thermodynamic modeling [J]. Process Safety Environmental Protection, **2016**, 102: 20-29.

[59] Hu Meng, Zheng Sunxiang, Mi Baoxia. Organic Fouling of Graphene Oxide Membranes and Its Implications for Membrane Fouling Control in Engineered Osmosis [J]. Environmental Science Technology, **2016**, 50 (2): 685.

[60] Wang Lin, Wang Naixin, Li Jie, et al. Layer-by-layer self-assembly of polycation/GO nanofiltration membrane with enhanced stability and fouling resistance [J]. Separation Purification Technology, **2016**, 160: 123-131.

[61] Igbinigun Efosa, Liu Yaolin, Ramamoorthy Malaisamy, et al. Graphene oxide functionalized polyethersulfone membrane to reduce organic fouling [J]. Journal of Membrane Science, **2016**, 514: 518-526.

[62] Oh Yoontaek, Armstrong Dana L, Finnerty Casey, et al. Understanding the pH-responsive behavior of graphene oxide membrane in removing ions and organic micropollulants [J]. Journal of Membrane Science, **2017**, 541: 235-243.

[63] Armstrong Dana L., Rice Clifford P., Ramirez Mark, et al. Influence of thermal hydrolysis-anaerobic digestion treatment of wastewater solids on concentrations of triclosan, triclocarban, and their transformation products in biosolids [J]. Chemosphere, **2017**, 171: 609-616.

[64] E Carey Daniel, Zitomer Daniel, Hristova Krassimira, et al. Triclocarban Influences Antibiotic Resistance and Alters Anaerobic Digester Microbial Community Structure [J]. Environmental Science Technology, **2015**, 50 (1): 126-134.

[65] Delgado Daniel R, Holguin Andres R, Martinez Fleming. Solution thermodynamics of triclosan and triclocarban in some volatile organic solvents [J]. Vitae, **2012**, 19 (1): 79-92.

[66] Liu Fei-fei, Zhao Jian, Wang Shuguang, et al. Effects of solution chemistry on adsorption of selected pharmaceuticals and personal care products (PPCPs) by graphenes and carbon nanotubes [J]. Environmental science technology, **2014**, 48 (22): 13197-13206.

[67] Pei Zhiguo, Li Lingyun, Sun Lixiang, et al. Adsorption characteristics of 1, 2, 4-trichlorbenzene, 2, 4, 6-trichlorophenol, 2-naphthol and naphthalene on graphene and graphene oxide [J]. Carbon, **2013**, 51 (1): 156-163.

[68] Xu Jing, Wang Li, Zhu Yongfa. Decontamination of bisphenol A from aqueous solution by graphene adsorption [J]. Langmuir, **2012**, 28 (22): 8418-8425.

[69] Maliyekkal Shihabudheen, Sreenivasan Sreeprasad, Krishnan Deepti, et al. Graphene: a reusable substrate for unprecedented adsorption of pesticides [J]. Small, **2013**, 9 (2): 273-283.

[70] Yu Baowei, Bai Yitong, Ming Zhu, et al. Adsorption behaviors of tetracycline on magnetic graphene oxide sponge [J]. Materials Chemistry Physics, **2017**, 198: 283-290.

第9章
CHAPTER 9

氧化石墨烯在光电传感领域的应用

　　石墨烯在光子和光电子领域与别的材料相比具有诸多优点[1]。作为零带隙材料，石墨烯的光响应谱覆盖了从紫外到 THz 范围；石墨烯在室温下就有着很高的电子迁移速度，这使得光子能量转换为电流或电压输出的响应较快[2,3]。石墨烯的低耗散率以及通过结构调控把电磁场能量限定近场区域的性质，使得光与石墨烯具有很强的相互作用[4,5]。

　　GO 和 RGO 由于官能团和缺陷的存在，较难观测到本征石墨烯中的一些电子输运效应以及其他物理效应，但其易于规模化制备、性质可调等优异特性，使其在传感检测领域展现出应用前景。RGO 在边缘处和面内缺陷处具有丰富的分子结合位点，使其成为一种很有希望的电化学传感器材料。结合原位还原技术，有很多研究使用诸如喷涂、旋涂等基于溶液的技术手段，利用 GO 在不同基底上制造出具备石墨烯相关性质的器件，以期在一些场合替代 CVD 法等制备的石墨烯薄膜。

　　GO 本身的电学性质与 sp^3 杂化和 sp^2 杂化的相对比例息息相关[6]，调节含氧基团相对含量可以实现 GO 从绝缘体到半导体再到半金属性质的转换[7,8]。此外，GO 在很宽的光谱范围内具有光致发光性质，同时也是高效的荧光淬灭剂。GO 特殊的光学性质和多样化的可修饰性，为石墨烯在光学、光电子学领域的应用提供了一个功能可调控的强大平台[6]，其在光电领域的应用日趋广泛。

　　GO 和 RGO 应用于光电传感，可以作为电子给体或者电子受体材料。本章在介绍其在光电领域的应用之前，对 GO 和 RGO 的相关光学性质部分进行介绍。

9.1 氧化石墨烯和还原氧化石墨烯的光学性质

GO 的光学性质与石墨烯有着很大差别。石墨烯本质上是零带隙半导体,在可见光范围内光吸收系数近乎常数(约 2.3%);相比之下,GO 的光吸收系数要小一个数量级(约 0.3%)[9,10]。而且,GO 的光吸收是波长的函数,其吸收曲线峰值在可见光与紫外光交界附近,随着波长向近红外一端移动,吸收系数逐渐下降。对紫外光的吸收(200~320nm)会表现出明显的 π-π^* 和 n-π^* 跃迁,而且其强度会随着含氧基团的出现而增加[11]。GO 的光响应对其含氧基团十分敏感[12]。随着含氧基团的去除,GO 在可见光波段的光吸收率迅速上升,最终可达到 2.3%[9,13]。GO 能级结构的理论和实验研究均指出,在还原过程中由于带间和带内光学跃迁机制的变化,GO 在还原过程中呈现不同的光吸收特性,利用这一特性可以大范围调控石墨烯的带隙宽度[14,15]。文献[16]中给出了计算得出的 GO 还原过程中能带值的变化。

GO 在吸收紫外/可见光后会发出荧光[11,17~20]。通常可以在可见光波段观测到两个峰值,一个在蓝光段(400~500nm),另一个在红光段(600~700nm)[6,17]。然而,对于 GO 发射荧光的机理,学界仍有争论。此外,氧化石墨烯的荧光发射会随着还原的进行逐渐变化,文献[16]在轻度化学还原过程中观察到 GO 光致发光光谱发生红移,这一发现与其他人观察到的发生蓝移的现象相矛盾[21,22],这从另一个方面说明了 GO 结构的复杂性和性质的多样性。

GO 中,碳的主要形式包含 sp^2 碳(石墨)和 sp^3 碳(氧化官能团)两种[23]。sp^3 碳区域将 GO 分割成了不同尺寸、不同形状的 sp^2 碳团簇,并将 sp^2 碳团簇团团包住,使其局域化(这种结构已在 GO 的透射电镜的研究中得到证明)。这里主要介绍 GO 的光学性质包括光的吸收、光致发光、荧光淬灭和超快光学非线性等几个方面。

9.1.1 氧化石墨烯的光吸收特性

紫外——可见吸收光谱(Ultraviolet Visible Absorption Spectroscopy,简称 UV-Vis)是研究材料光学性质非常重要的方法。文献[16]给出了 GO 的紫外——可见吸收光谱。其中,在 227nm 处的吸收,其主要来自于电子在 π-π^* 之间的跃迁,300nm 处的吸收为电子从 n-π^* 间的跃迁[13,24]。研究发现,吸收光谱的带边非常弥散无法清晰分辨,说明 GO 内部分布着非常多的局域态。

300nm 处的吸收峰是 GO 中羰基存在的特征,XPS 的数据表明,GO 中存在的四种主要含氧官能团中,主要的是羟基和环氧基,羰基和羧基官能团相对较少[21,24~26]。

FTIR 也证实了 GO 中这些官能团的存在[26]。羟基、环氧和羧基等官能团的振动在 1500cm^{-1} 以下，而在 1500cm^{-1} 以上大约 1700cm^{-1} 处存在一个可探测的羰基 C=O 伸缩振动产生的峰[19,27,28]。

此外，GO 周围介质 pH 值的变化也会对其紫外——可见吸收光谱产生一些影响，但不影响峰的绝对强度以及它们的位置[13,19]。然而，红外吸收光谱却显示出官能团在不同 pH 值条件下质子化/去质子化有显著变化[26]。文献 [26] 给出了在酸性和碱性条件下 GO 的红外吸收光谱，显示出跃迁能量存在显著差异，这一差异来自于酸性或碱性环境中官能团氧化状态的变化。基于这一特性，红外吸收光谱可以作为研究 GO 中官能团修饰改性状态的有力工具。

9.1.2 氧化石墨烯的发光特性

官能团的存在打开了带隙，因此 GO 具备半导体光致发光的特性。早在 19 世纪中叶 Brodie 就研究了通过石墨化学氧化产物中的光学性质[29]，Hummers 在 1958 年研究了石墨氧化物材料中的带隙[29,30]。研究者曾经在 GO 的水悬浮液中观察到两种完全不同类型的光致发光现象。一些研究发现在离心分离纯化的少层 GO 中观察到蓝色（350~450nm）的光致发光[11]，这可以归因于氧化引入的 sp^3 结构分隔了 sp^2 石墨区域中的电子—空穴复合。同时，另外一些研究人员发现化学氧化法得到的 GO，其光致发光光谱却在绿色到红外（500~800nm）波段[19,20,31,32]。普遍认为，这个波段的光致发光现象是由局域电子结构、周围官能团[19,20]、sp^3 碳势垒包围的 sp^2 碳区域或者化学键变化导致的谷间散射效应引起的[32]。Eda 等[11] 人认为，红光波段荧光发射是多层石墨烯片层间弛豫过程红移的结果。

GO 的光致发光强度会随着温度的变化而变化[34]。相比之下，非晶碳薄膜的发光强度几乎不随温度的升高而发生改变。研究表明，在非晶碳材料体系中，即使在具有较高的缺陷密度时，其发光强度仍然受温度影响较小[35]。这是由于光激发的电子—空穴对被 sp^3 碳高势垒区限制在同一个碳团簇中，导致电子、空穴的波函数交叠变大，大大增加了发生辐射跃迁的概率[35,36]。而当 sp^3 势垒层被破坏时，在声子辅助下电子很快跑到缺陷处复合，导致其在室温下很难观察到发光现象。GO 的发光随温度的变化介于两者之间，这是由 GO 的结构和成分组成决定的，即一方面存在高势垒 sp^3 区域对 sp^2 碳团簇的限制，增大了辐射跃迁概率，另一方面由于氧化官能团引入的高能振动以及 GO 的柔性结构增强了电子和声子的相互作用，而增大了无辐射跃迁的概率。

其他研究表明，GO 的发光来自于不同的局域状态，而不同能量的激发光又对带隙与其能量相对应的 sp^2 碳团簇具有共振效应，因此发射峰的中心波长跟随激发光的能量减小而红移；激发光能量的降低，将导致被激发的局域状态数目减少，因此发射峰变窄。

9.1.3 还原氧化石墨烯的光吸收特性

GO 在还原成 RGO 的过程中，光吸收行为发生了巨大的变化。GO 被还原后，电子的 π-π^* 吸收峰从 230nm 移动到了 270nm，这是由于含氧官能团被去除，π 共轭体系拓展，π-π^* 间带隙变窄的结果[26,37]。同时光的吸收在整个可见波段明显增强，这也是由于 π 自由电子的恢复所致。另外，由于含氧官能团的消失，300nm 处电子从 n-π^* 间跃迁的吸收基本消失。以上吸收光谱的变化是 GO 发生还原反应的重要标志。

对吸收光谱进行更细致地分析，可以从中得到更多关于 RGO 结构的信息。GO 经历还原处理的过程中，sp^2 碳区域拓展并最终连接在一起，导致了带隙变窄以及局域态数目减少。

由于热还原过程存在分解，氧化官能团转化成了 CO 和 CO_2 的同时将引入大量的碳空位以及碳缺陷，这增加了体系的无序度[38]；而化学还原则可以修复 GO 中的苯环结构，将改善体系的无序化程度。化学还原和热还原的反应机理不同，导致了两者还原效果的差异。可以通过 GO 还原后吸收光谱和拉曼光谱的对比研究，判断不同还原方法对 GO 结构的影响，这将对设计和开发有效的还原方法具有重要意义。

9.1.4 还原氧化石墨烯的发光特性

GO 被完全还原后，氧化官能团将被除掉，sp^2 碳团簇拓展并最终连接起来，构成了大的 π 共轭体系。GO 逐渐向石墨烯结构转变，发光特性将消失。

文献[38]给出了一个反向研究过程，即使用臭氧对 RGO 进行氧化得到 GO，在此过程中吸收光谱出现了与还原过程完全相反的变化：随着氧化时间的增加，吸收光谱的峰值向蓝光方向移动，可见光的吸收显著降低。这表明在氧化过程中 sp^2 石墨结构逐渐减少，而 sp^3 区域逐渐形成。因此，当讨论 RGO 的发光特性时，一般与 RGO 被还原的程度相关。这种可见光的吸收光谱变化特征，可以作为一种表征技术，用于氧化或还原中 GO 中含氧官能团含氧量的变化研究。

9.1.5 氧化石墨烯量子点

与石墨烯量子点类似，氧化石墨烯量子点也具备一些特殊的性质。当 GO 片径小至若干纳米量级的时候将会出现明显的限域效应，其光学性质会随着片径尺寸大小发生变化[39]，当超过某上限后氧化石墨烯量子点的性质相当接近 GO。这就提供了一种通过控制片径尺寸分布改变氧化石墨烯量子点光响应的手段。以 GO 为前驱体通过超声——水热法得到的石墨烯量子点的光发射性能，在蓝光区域其光发射性能取决于 zigzag 边缘状态，而绿色的荧光发射则来自于能级陷阱的无序状态。通过控制氧化石墨烯量子点的氧化程度，可以控制其发光的波长。

对于片径小于 5nm 的量子点，研究表明，依赖激发的荧光效应并不是由 GO 中广泛存在的 sp^2 石墨区域造成，而更倾向于是由于量子点本身的分布造成的[39]。这些 GO 量子点的荧光量子产率可以达到 11.4%，这一数值要远高于 GO 的量子产率。

一般来说，石墨烯量子点的荧光效应都来自于含氧基团在低维结构上的出现，而 GO 展现出的依赖激发光波长的发射特性，则是一些综合因素造成的，包括表面缺陷、限域效应以及边缘形状等。

9.1.6 荧光淬灭

尽管 GO 自身可以发射荧光，但有趣的是它也可以淬灭荧光。这两种看似相互矛盾的性质集于一身，正是由于 GO 化学成分的多样性、原子和电子层面的复杂结构造成的。众所周知，石墨形态的碳材料可以淬灭处于其表面的染料分子的荧光，同样的，在 GO 和 RGO 中存在的 sp^2 区域可以淬灭临近一些物质的荧光，如染料分子、共轭聚合物、量子点等，而 GO 的荧光淬灭效率在还原后还有进一步的提升。相关研究给出了 GO 淬灭荧光的机理，指出由于染料分子发射的荧光光谱范围与 GO 的光吸收谱范围没有重叠，因此 GO 的荧光淬灭实际上是一个电子的转移过程[40]。而染料分子与 GO 间的能量传递效率，与染料分子与 GO 间的结合能有关，因此通过调控 GO 表面的官能团，可以利用这种效应作为荧光淬灭显微技术（Fluorescence Quenching Microscopy，简称FQM）的基础[41]。这种显微技术构建的方法与实际效果也得到了报道[42]。基于 FQM 的技术使用了与通过激发使得目标在黑暗背景下变亮相反的方法，即在目标及其周边地区上覆盖一层荧光层，激发时，石墨烯基片材在明亮背景下显得黑暗。二维材料的长程荧光淬灭能力也使得区分重叠层、皱纹和褶皱成为可能，因为它们会比单层显得更暗。因此，FQM 能够产生具有对比度和层分辨率的清晰图像，和从 AFM 和 SEM 中获得的图像在一个水平上。FQM 的另一个优势是其在不同格式下的灵活性：能够在任意衬底和溶液中实现二维片材成像。FQM 显著提高了 GO 和 RGO 的对比度，与通常依赖于干扰效应和需要特殊设计衬底的常规光学成像技术相比，提高了对比度。在荧光淬灭效应的基础上，石墨烯被用作衬底来抑制共振拉曼光谱中的荧光干扰，以产生相对信号增强。

9.1.7 光学非线性

光学材料的某些非线性性质是实现高性能集成光子器件的关键。光子芯片的许多重要功能，如全光开关、信号再生、超快通信都离不开它。找寻一种具有超高三阶非线性并且易于加工成各种功能性微纳结构的材料是众多的光学科研工作者的目标。

超快泵浦探针光谱表明，重度功能化的具有较大 sp^3 区域的 GO 材料在高激发强度下可以出现饱和吸收、双光子吸收和多光子吸收[6,43~45]，这种效应归因于在 sp^3 结构域中存在较大的带隙。相反，在具有较小带隙的 sp^2 域中的仅出现单光子吸收。石墨烯在飞秒脉冲激发下具有饱和吸收[45]，而 GO 在低能量下为饱和吸收，高能量下则具有反饱和吸收[44]。因此，通过控制 GO 氧化/还原的程度，实现 sp^2 域到 sp^3 域的比例调控，可以调整 GO 的非线性光学性质，这对于高次谐波的产生与应用是非常重要的。

在 GO 还原成 RGO 的过程中，材料的导电性、禁带特性和折射率都会发生连续变化，形成独特而优异的可调谐型新材料。2014 年，澳大利亚微光子学中心贾宝华等发现在用激光直写氧化石墨烯薄膜形成微纳米结构的过程中，材料的非线性可以实现激光功率可控的动态调谐[46]。

与传统的非线性材料相比，GO 的三阶非线性高出了整整 1000 倍，随着 GO 中的氧成分逐渐减少，而非线性也呈现出被动态调谐的丰富变化。不但材料的非线性系数的大小产生改变，其非线性吸收和折射率也发生变化，并且，这种丰富的非线性特性完全可以实现动态操控。

从实现功能器件的角度来说，非线性品质因子是一个非常重要的参数。非线性品质因子是否大于 1，普遍被认为是能否实现可集成化非线性器件的判据。传统的非线性材料非线性品质因子都远远小于 1，用这些材料形成的功能器件将需要很长的作用距离才能达到所需的非线性调制。因此器件的小型化、集成化面临巨大挑战。GO 薄膜的非线性品质因子高达 4.56，属于已知非线性材料中最高的一类。

9.2 光电传感领域中的氧化石墨烯

光电器件是在微电子技术基础上发展起来的一种实现光与电之间相互转换的器件，其核心是各种光电材料，即能够实现光电信息的接收、传输、转换、监测、存储、调制、处理和显示等功能的材料。光电传感器件指的是能够对某种特征量进行感知或探测的光电器件，狭义上仅指可将特征光信号转换为电信号进行探测的器件，广义而言，任何可将被测对象的特征转换为相应光信号的变化、并将光信号转换为电信号进行检测、探测的器件都可称之为光电传感器。

GO 同时具有荧光发射和荧光淬灭特性，广义而言，其自身已经可以作为一种传感材料，在生物、医学领域的应用充分说明了这一点。经过功能化的 GO/RGO 在更加广泛的领域内得到了应用，特别在光探测、光学成像、新型光源、非线性器件等光电传感相关领域有着丰富的应用。

9.2.1 氧化石墨烯——光探测器件

光电探测器是石墨烯问世后最早应用的领域之一。2009 年，Xia 等[2] 人利用机械剥离的石墨烯制备出了石墨烯光电探测器。该研究以1～3层石墨烯作为有源层，Ti/Pd/Au作源漏电极，Si作为背栅极并在其上沉淀300nm厚的SiO_2。在电极和石墨烯的接触面上因为功函数的不同，能带会发生弯曲并产生内建电场。当光照射在内建电场区域时，石墨烯价带中的电子受激跃迁到导带，产生的电子空穴对在内建电场的作用下分离，向源漏极输运产生光电流。但是当光照射内建电场以外的地方时，在石墨烯上产生的电子空穴对会很快复合，对漏电流没有贡献。由于其内建电场范围只有0.2nm，因此获得的光电响应只有0.5mA/W。虽然其光电响应很低，但其频带宽度可以达到40GHz，而光电流没有下降，而且其频带宽度有望达到500GHz。

经过多年的不断发展，石墨烯光电探测器的种类也不断丰富。依据结构与原理的差异，基于石墨烯的光电探测器主要有金属-石墨烯-金属型、光电导型、半导体异质结型及其他一些新型结构探测器。早期基于石墨烯的光探测器均采用的是CVD法或者机械剥离石墨烯，其制备、转移工艺难度较大。GO因其易于制备、组装、修饰，并且可通过控制还原程度调控其性质，在近些年得到了广泛的研究。

2014年，Liang等[47]报道了RGO的中红外光响应效应，其响应率约为1A/W。最近，基于石墨烯的红外光电探测器被证明具有高达40GHz的光调制响应能力，但光响应度较低（约6.1mA/W），这是由光生载流子快速复合造成的；2013年，Chang等[48]人通过控制缺陷和原子结构制备了基于RGO的红外光电晶体管，光响应达到了约0.7A/W（895nm），外部量子效率达到了97%[48]。尽管石墨烯不存在带隙，在金属-石墨烯界面仍然观测到了较强的光电流响应，其内部量子效率为15%～30%[49]。Liu等[50]人利用石墨烯双层异质结设计了宽带光探测器。2017年，Liu等[51]人研制了一种常温下具有高达105A/W超高光响应率的硅-石墨烯光电导电探测器。

以上这些光探测器不同的性能，源自于不同的石墨烯合成路线、方法和器件制造工艺。下面介绍几种近年来研究较多的基于石墨烯的光探测器。

（1）金属接触型光探测器（Metal-Graphene-Metal，简称MGM）

金属-石墨烯-金属型光探测器是石墨烯光探测器中研究最早的一种类型[52,53]。早期的此类探测器中使用的是CVD法石墨烯，石墨烯在其中连接源端和漏端，起到沟道作用。该类探测器借鉴了传统的场效应晶体管结构，优势是具有宽光谱响应，响应电流可通过调节门电压进行调控。其结构与现有微电子器件和电路有很好的兼容性。

2010年，Thomas等[3]人提出了一个改进的结构，不对称的叉指结构的石墨烯探测器（Metal-Graphene-Metal Photodetectors，简称MGMPD）。该结构利用两种不同功函数的金属作源漏电极。当使用相同的金属作电极时，光照射在电极附近，金属和石墨烯的接触面上产生大小相等方向相反的内电场，会导致等大反向的光电流，使得总光电流为零。采用不对称电极结构（钛、钯），使无源漏偏压时净光电流不为零，得到了 1.5mA/W 的光响应率，16GHz 的 3dB 带宽。在外加源陋偏压时，得到了最大光响应 6.1mA/W。

在以上工作基础上，2015 年，Jabbarzadeh 等[54]人使用水合肼还原 GO 替代石墨烯，制备了"金属-还原氧化石墨烯-金属"结构的中红外光探测器。研究中对不同还原程度的 RGO 制备的光探测器性能进行了对比，得到了基于强还原 GO 的高灵敏度、高响应度的中红外探测器。

2017 年，Mustaque 等[55]人使用溶剂热法还原氧化石墨烯，制备了一种"金属-还原氧化石墨烯-金属"结构的近红外光探测器，其光响应达到了 1.34A/W，外部量子效率＞100%，响应时间在若干秒量级。据称，这是当时单独使用 RGO 作为光敏材料可以获得的最大光响应度。

"金属-还原氧化石墨烯-金属"结构的光探测器有着较高的响应度，但是其缺点是响应时间稍长。

（2）半导体异质结型光探测器

石墨烯可与多种传统半导体材料形成异质结，如硅[56~58]、锗[59]、硒化镉[60]、硫化镉[61]、二硫化钼[62]等。其中，石墨烯/硅异质结器件是目前研究最为广泛、光电转换效率最高（AM1.5）的一类光电器件。

基于硅-石墨烯异质结光电探测器，获得了极高的光伏响应[63]，并有极小的等效噪声功率可以探测极微弱的信号；相比于光电流响应，它不会因产生焦耳热而产生损耗。

RGO 制备简单、自身具有受还原程度调控的带隙，可以实现超宽谱（从可见至太赫兹波段）探测。GO 的还原程度对探测性能有显著影响。随着 GO 还原程度的提高，探测器的响应率可以提高若干倍以上。因此，在 CVD 石墨烯研究方案的基础上，研究者开始尝试使用 RGO 制备类似结构的光电探测器。对于 RGO-Si 器件，带间光子跃迁以及界面处的表面电荷积累，是影响光响应的重要因素[64]。

2014 年，Cao 等[65]人将 GO 分散液滴涂在硅线阵列上，而后通过热处理对 GO 进行热还原，制得了硅纳米线阵列（SiNW）-RGO 异质结的室温超宽谱光探测器。该探测器在室温下，首次实现了从可见光（532nm）到太赫兹波（2.52THz，118.8mm）的超宽谱光探测。在所有波段中，探测器对 10.6mm 的长波红外具有最高的光响应率可达 9mA/W。

Singh 等[66]人于 2018 年采用背靠背连接的 np-pn 结构，制备了基于大带隙 RGO 的无金属宽谱（300nm 至 1100nm）混合光电探测器，得到的 n-RGO/p-Si 异质结在近红外区（830nm）实现了在控制光照条件；1V 偏压、光照 $11\mu W/cm^2$ 下，峰值响应率 16.7A/W，峰值探测率为 2.56×10^{12} 琼斯，响应速度 $460\mu s$，恢复时间 $446\mu s$ 的优异性能，并且制备的器件具有良好的重复性、开关特性、耐用性和极高的外部量子效率（2.5×10^3%），以及在弱光条件下的超灵敏特性。

(3) 肖特基结型光探测器

肖特基结光电探测器具有较简单的结构,具备暗电流小、响应速度快、寄生电容小等优点。越来越多的研究表明,RGO 材料在 Ni-Si 肖特基结光电探测器的构建中能有效替代 CVD 法和机械剥离石墨烯,所得器件具有良好的光电特性,如开关比、响应度和探测性能。RGO 材料中的缺陷会影响器件的各种特性,改变 RGO 的还原水平,可以对诸多参数在零偏压条件下进行调节,但几乎不影响器件的开关比或探测率[67]。

Guo 等[68]人开发了一种基于光电肖特基结的快速、超灵敏的人工嗅觉系统,可实现对不同爆炸物蒸汽的识别性检测,包括 TNT、DNT、PNT、PA 等。这个方法主要基于光电子 SiNWs/ZnO/rGO 肖特基传感器,这较传统的肖特基传感器,在夜间工作时,对爆炸物蒸汽更是超灵敏、快速光电响应和光强度可调。

Dehkharghani 等[69]人于 2017 年制备的一种基于 RGO 的新型肖特基红外探测器表明,RGO 可以与 Si 形成整流结,在相同的外部反偏电压(12 V)下,该探测器的响应度约为 2.021A/W,比传统商用标准器件(例如滨松 StopyLtd 有限公司 S6967 型)高出 3 倍。同时,这种装置也可以产生光伏效应,意味着该结构的探测器在太阳能电池和低电压光电检测设备中也有可能得到应用。

(4) 自支撑型光探测器

基于 RGO 的自支撑型光探测器,充分发挥了 RGO 相对于 CVD 法和机械剥离石墨烯的优势,即可以组装成为大面积薄膜、光谱响应范围宽、性能可调控等优点。

Yang 等[70]人将分散在乙醇中的 GO 通过交替滴注和干燥,层层组装到预先准备衬底上,然后在高温下退火,制备具有极高电导率的 RGO 薄膜。所制备的 RGO 薄膜电导率达到了 87100S/m,面电阻 21.2Ω/□,迁移率 16.7cm^2/(V·s)。在当时所有已报道的 RGO 薄膜光电探测器中实现了最快(约 100 ms)和最宽(从紫外光到太赫兹光谱范围)的光谱响应,光响应速度可以与 CVD 生长的石墨烯光电探测器和机械剥离石墨烯光电探测器相媲美。

Abid 等[71]人研究了基于 RGO 的自支撑型光探测器其温度、缺陷以及量子效率之间的关系。采用改进的 Hummers' 方法制备 GO,并在氩气气氛中 250℃下热还原 1 小时获得 RGO,通过真空过滤制备自支撑 RGO 膜,进而制成了光探测器。所开发的传感器对于 635nm 的照明波长表现出最高的灵敏度,和工作温度相关度不大。对于给定的激发波长,在低温下(123~303K)灵敏度和工作温度之间呈反比关系。在功率密度为 1.4mW/mm^2 的 635nm 激光、123 K 温度条件下,灵敏度最高达到 49.2%。

9.2.2 氧化石墨烯作为新型光源材料

(1) 脉冲激光器

短脉冲、超短脉冲激光器由于在核聚变、光谱分析、高速成像、激光雷达、激光测距等领域有着十分重要的应用，一直是光电领域研究的热点之一。调 Q 技术与锁模技术是应人们对高峰值功率、窄脉宽激光脉冲的应用需求而发展起来的。两种方式机理不同，压缩的程度也不同。调 Q 技术可将激光脉宽压缩至纳秒量级（峰值功率达 10^6 W），锁模技术可将激光脉宽压缩至皮秒或飞秒量级（峰值功率可达到 10^{12} W）。调 Q 技术与锁模技术的出现和发展，是激光发展史上的一个重要突破。调 Q 和被动锁模都可以采用可饱和吸收体的光学非线性特性，都可以产生短脉冲。只不过调 Q 机制中，可饱和吸收体的恢复时间比脉冲宽度大，在锁模中可饱和吸收体的恢复时间比脉冲宽度短，可以产生超短脉冲。

目前，比较成熟的非线性材料有半导体可饱和吸收镜和碳纳米管可饱和吸收体。但是制作半导体可饱和吸收镜需要相对复杂和昂贵的超净制造系统，这类器件的典型恢复时间约为几个纳秒，且半导体可饱和吸收镜的光损伤阈值很低，常用的半导体饱和吸收镜吸收带宽较窄。碳纳米管是一种直接带隙材料，带隙大小由碳纳米管直径和属性决定。不同直径碳纳米管的混合可实现宽的非线性吸收带，覆盖常用的 $1.0\sim1.6\mu m$ 激光增益发射波段。但是由于碳纳米管的管状形态会产生很大的散射损耗，提高了锁模阈值，限制了激光输出功率和效率，所以，研究人员一直在寻找一种具有高光损伤阈值、超快恢复时间、宽带宽和价格便宜等优点的饱和吸收材料。

近年来，研究者发现石墨烯由于它独特的零带隙结构，对所有波段的光都无选择性的吸收，且具有超快的恢复时间和较高的损伤阈值。因此利用石墨烯独特的非线性可饱和吸收特性将其制作成可饱和吸收体应用于调 Q 掺铒光纤激光器、被动锁模光纤激光器已经成为超快脉冲激光器研究领域的热点。2009 年，Bao 等[72] 人使用单层石墨烯作为锁模光纤激光器的可饱和吸收体首先实现了通信波段的超短孤子脉冲输出，脉冲宽度达到了 756fs。他们证实了由于泡利阻塞原理，零带隙材料石墨烯在强激光激发下可以容易的实现可饱和吸收，而且这种可饱和吸收是与频率不相关的，即石墨烯作为可饱和吸收体可实现对所有波长的光都有可饱和吸收作用。2010 年，厦门大学报道了石墨烯调 Q 掺铒光纤激光器[73]（重复频率 $3.3\sim65.0$ kHz，脉宽 $3.7\mu s$，单脉冲能量 16.7nJ）。2011 年，剑桥大学报道了中心波长可调的石墨烯调 Q 光纤激光器[74]（重复频率 $36\sim103$ kHz，脉宽为 $2\mu s$，单脉冲能量 40nJ）。

由于 CVD 法和机械剥离石墨烯的一些挑战，近些年来人们已经开始探索将 GO 及其复合材料应用于这一领域。Wang 等[75] 人于 2011 年用一种低成本氧化石墨烯"壁纸"方案制备成吸收体，实现了被动调 Q 和锁模 Nd：YVO4 激光器。这种吸收器由高透射镜、氧化石墨烯壁纸构成吸收体和反射镜，成功地实现了平均 310MW 功率的被动调 Q 和锁模激光器输出。重复 Q 开关包络的速率和脉冲宽度分别为 213kHz 和 770ns。在 Q 开关包络中，被动锁模脉冲的重复频率为 81.3MHz，脉冲能量为 3.8nJ。这种结构简单、易于制备，但是吸收介质损伤阈值较低，长时间运行出现了介质损伤现象。

Yu等[76]人于2013年制备了一种使用GO作为吸收体的瓦级被动调Q掺镱双包层光纤激光器。GO型可饱和吸收镜（GO-SAM）的结构为三明治夹层型结构，在1044nm波长附近获得最大输出功率1.8W。据称，这是当时基于GO饱和吸收体的Q开关光纤激光器可达到的最高输出功率。GO被保护在气体中，从而有效地提高了GO-SAM的损伤阈值。增益纤维为D型掺镱双包层光纤。脉冲重复频率从120到215kHz，泵浦功率从3.89W到7.8W，最大脉冲能量为8.37μJ，脉冲宽度为1.7μs。

Chang等[77]人在2017年使用改进的Hummers'法制备了少层氧化石墨烯饱和吸收器件（记作GO-SA），其中所制备的GO富含羟基，这有利于在衬底上成膜以制造Q开关元件。实验结果表明，GO-SA是被动调Q绿光激光器的一种有前途的饱和吸收体。所制备的Nd：YVO4/PPLN绿光激光器，在最大泵浦功率为5.16W时，得到了最大平均输出功率536mW，重复频率、脉冲宽度和中心波长分别为71.4kHz、98ns和531.2nm。可以估算出7.6nJ的单脉冲能量和766W的峰值功率。GO有望用作被动调Q激光器的高性能SA、光子器件的非线性光学材料、光开关等。

（2）THz光源

THz波（太赫兹波）包含了频率为0.1到10THz的电磁波。在20世纪80年代中期之前，太赫兹波段两侧的红外和微波技术发展相对比较成熟，但是人们对太赫兹波段的认识仍然非常有限，形成了所谓的"THz Gap"。太赫兹技术可用于医学诊断与成像、反恐安全检查、通信雷达、射电天文等领域，将对技术创新、国民经济发展以及国家军事安全等领域产生深远的影响。作为极具发展潜力的新技术，2004年，美国政府将THz科技评为"改变未来世界的十大技术"之一，而日本于2005年1月8日更是将THz技术列为"国家支柱十大重点战略目标"之首。传统的宽带THz波可以通过光整流、光电导天线、激光气体等离子体等方法产生，窄带THz波可以通过太赫兹激光器、光学混频、加速电子、光参量转换等方法产生。

基于石墨烯的THz研究在2007年以前就开展了。Ryzhii等[78~83]人研究了光抽运下石墨烯负动态电导率的实现机理理论证明了利用石墨烯产生THz辐射的可行性。日本东北大学与会津大学在2009年的合作研究中发现，在硅基板上制作石墨烯薄膜，通过红外激光的照射，只需短暂的时间在材料表面就能产生太赫兹光[80]。2011年，Karasawa等[82]人首次获得光抽运单层石墨烯产生太赫兹辐射的实验结果。

2017年，Wang等[84]人采用改进的Hummers'方法制备了GO，采用真空抽滤法制备了不同厚度的GO薄膜，并通过热退火获得了不同还原程度的RGO薄膜。飞秒激光脉冲入射到入射角为45°的样品，用反射角45°测量THz波。实验表明，通过促进还原程度可以提高THz辐射的发射，这与sp^2杂化碳畴比例有关。

同时发现，RGO 的 THz 生成效率随着薄膜厚度的增加而下降，这可以归因于 RGO 表面的光诱导横向电流受到了抑制。当激发激光从样品反面一侧照射时，可以观察到产生的 THz 波产生了相移，这表明 RGO 的 THz 生成的机制受光子牵引效应影响。基于这一理论，实验测量的 THz 电场的偏振角依赖性与理论计算一致。这一工作明确了 RGO 的 THz 生成机制，为室温下开发新型 THz 源奠定了基础。

9.2.3 氧化石墨烯-光纤传感器件

在光通信领域，Xu 等人[85] 开发了飞秒氧化石墨烯锁模掺铒光纤激光器，与基于石墨烯的可饱和吸收体相比，性能有所提升并且具有易于制造的优点，这是 GO/RGO 在与光纤结合应用最早的报道之一。在传感领域，Sridevi 等[86] 人提出了一种基于腐蚀布拉格光栅光纤外加 GO 涂层的高灵敏、高精度生化传感器，该方法在检测刀豆球蛋白 A 中进行了试验。

为了探索光纤技术和 GO 特性结合的优点，文献 [89] 介绍了不同的 GO 涂层在光纤样品上应用的特点，还分析了在倾斜布拉格光栅光纤表面增加 GO 涂层对折射率变化的影响，论证了这种构型对新传感器的发展的适用性。

Xiao 等[88] 人于 2014 年首次制备了基于 RGO 的光纤湿度传感器。采用 RGO 乙醇悬浮液，通过随机沉积法在侧抛光光纤的抛光表面上涂覆 RGO 膜。侧抛光光纤由普通纤维制成，从一侧抛光熔覆层的一部分，抛光表面的覆盖材料与倏逝波之间的相互作用将调制纤芯内部的光场用于制作传感器。所制备的传感器在 70%～95% Rh 内实现功率变化达 6.9dB，线性响应系数为 98.2%，灵敏度为 0.31dB/%RH，响应速度快于 0.13% RH/s，75%～95% Rh 具有良好的重复性。石墨烯膜侧抛光光纤传感器在其他化学气相检测中也有广阔应用前景。依赖于新颖的传感机制，石墨烯膜侧抛光光纤可以作为现有石墨烯基电化学气敏传感器的有益补充。这种新型石墨烯光学传感器不仅有助于克服电学上的缺点，而且还具有制造成本低、遥感能力强、抗电磁干扰、多路复用等优点。

2014 年，Gupta 等[89] 人用 RGO 制备了 Cu/PMMA/RGO 复合材料作为光纤等离子体共振传感器的传感层，对多种常见气体进行了检测，其中对氨气具有优异的选择性，该传感器具有优秀的可重复性，具有环境监测的应用价值。

2014 年，Yaacob 等[90] 人把锥形光纤探针与 GO 相结合制备成酒精浓度光纤传感器，利用 GO 锥形光纤传感器的反射响应对 5%～80% 体积浓度的酒精溶液进行了检测，结果表明，组装了 GO 的锥形光纤传感器比裸的锥形光纤传感器的灵敏度提高很多。

2016 年，Rao 等[91] 人在单模光纤拉锥区域组装了铂纳米颗粒修饰的 GO 薄膜，对浓度为 0～120ppm 的氨气进行了检测，实验结果证明，有铂纳米颗粒修饰比没有铂纳米颗粒修饰的 GO 薄膜光纤传感器灵敏度高三倍，为多种气体的检测提供了一个理想的平台。

总之，GO/RGO 在光纤传感领域会有越来越多的应用，其基本的原理是利用石墨烯及氧化石墨烯的淬灭特性、分子吸附特性以及对金属纳米结构的惰性保护作用等，通过吸收光纤芯层穿透的倏逝波改变光纤折射率或者基于表面等离子体共振效应影响折射率。GO/RGO 可以在光纤的侧面、端面对光进行吸收或者反射，而为了增加光与 GO/RGO 层的相互作用，采用了不同光纤几何弯曲形状，如直型、U 型、锥型和双锥型等。

9.3 小结

本章介绍了基于氧化石墨烯/还原氧化石墨烯的传感器件，尤其是光探测器。作为广义的光电传感器件，任何可以在光电传感系统中用到的探测器、光源、传感器、调制器等，均可看作光电传感器件或其一部分，因此，本章除了介绍光探测器外，对新型光源和光纤传感器件也做了简单的介绍。

GO/RGO 在光电传感领域的应用，其基本依据是本章前面部分所涉及的各种光学性质。GO 因含氧官能团的存在具备了丰富的光学特性，在还原为 RGO 的过程中，不同的还原程度又具备了不同的性质，从结构方面而言，是其 sp^2 碳域与 sp^3 碳域相互分割、相互影响、相互转化带来了如此丰富的特性。也正是这些官能团的存在，使得 GO 可以方便地采用各种基于溶液的方法适应多种场合的需要，克服了 CVD 法和机械剥离石墨烯在转移和大面积应用时存在的缺点，也正是这些官能团的存在，使其便于实现功能化修饰，为其在不同场景的应用提供了一个广阔的平台。

── 参考文献 ──

[1] Bonaccorso Francesco，Sun Zicai，Hasan Tawfique，et al. Graphene photonics and optoelectronics [J]. Nature Photonics，2010，4（9）：611-622.

[2] Xia Fengnian，Mueller Thomas，Lin Yu-Ming，et al. Ultrafast graphene photodetector. in Lasers and Electro-Optics. 2009.

[3] Mueller Thomas，Xia Fengnian，Avouris Phaedon. Graphene photodetectors for high-speed optical communications [J]. Nature Photonics，**2010**，4（5）：297-301.

[4] Koppens Frank，E. Chang Darrick，Garcia de Abajo Javier. Graphene plasmonics：a platform for strong light-matter interactions [J]. Nano Letters，**2011**，11（8）：3370-3377.

[5] Grigorenko A. N.，Polini Marco，Novoselov K. S.. Graphene plasmonics [J]. Nature Photonics，**2012**，6（11）：749-758.

[6] Loh Kian Ping，Bao Qiaoliang，Eda Goki，et al. Graphene oxide as a chemically tunable platform for optical applications [J]. Nature Chemistry，**2010**，2（12）：1015-1024.

[7] Ito Jun，Nakamura Jun，Natori Akiko. Semiconducting nature of the oxygen-adsorbed graphene sheet [J]. Journal of Applied Physics，**2008**，103（11）：113712.

[8] Boukhvalov D. W., Katsnelson Mikhail I. Modeling of graphite oxide [J]. Journal of the American Chemical Society, **2008**, 130 (32): 10697-10701.

[9] Sokolov Denis A, Morozov Yurii V, McDonald Matthew P, et al. Direct observation of single layer graphene oxide reduction through spatially resolved, single sheet absorption/emission microscopy [J]. Nano letters, **2014**, 14 (6): 3172-3179.

[10] Zhou Yong, Bao Qiaoliang, Varghese Binni, et al. Microstructuring of Graphene Oxide Nanosheets Using Direct Laser Writing [J]. Advanced Materials, **2010**, 22 (1): 67-71.

[11] Eda Goki, Lin Yun-Yue, Mattevi Cecilia, et al. Blue photoluminescence from chemically derived graphene oxide [J]. Advanced Materials, **2010**, 22 (4): 505-509.

[12] Andryushina Natalya S., Stroyuk Oleksandr L., Dudarenko Galyna V., et al. Photopolymerization of acrylamide induced by colloidal graphene oxide [J]. Journal of Photochemistry Photobiology A: Chemistry, **2013**, 256 (4): 1-6.

[13] Li Dan, B Müller Marc, Gilje Scott, et al. Processable aqueous dispersions of graphene nanosheets [J]. Nature Nanotechnology, **2008**, 3 (2): 101-105.

[14] Huang Haiming, Li Zhibing, She Juncong, et al. Oxygen density dependent band gap of reduced graphene oxide [J]. Journal of Applied Physics, **2012**, 111 (5): 054317.

[15] Shen Yan, B. Yang S, Zhou Peng, et al. Evolution of the band-gap and optical properties of graphene oxide with controllable reduction level [J]. Carbon, **2013**, 62 (5): 157-164.

[16] Xin Guoqing, Meng Yinan, Ma Yifei, et al. Tunable photoluminescence of graphene oxide from near-ultraviolet to blue [J]. Materials Letters, **2012**, 74 (74): 71-73.

[17] Chien Chih-Tao, Li Shao-Sian, Lai Wei-Jung, et al. Tunable photoluminescence from graphene oxide [J]. Angewandte Chemie, **2012**, 51 (27): 6662-6666.

[18] Exarhos Annemarie, E Turk Michael, M Kikkawa James. Ultrafast spectral migration of photoluminescence in graphene oxide [J]. Nano Letters, **2013**, 13 (2): 344-349.

[19] Galande Charudatta, Mohite Aditya D., Naumov Anton V., et al. Quasi-Molecular Fluorescence from Graphene Oxide [J]. Scientific Reports, **2011**, 1 (6048): 85.

[20] Luo Zhengtang, Vora Patrick, Mele Eugene, et al. Photoluminescence and Band Gap Modulation in Graphene Oxide [J]. Applied Physics Letters, **2009**, 94 (11): 111909.

[21] Chen Jin Long, Yan Xiu Ping. A dehydration and stabilizer-free approach to production of stable water dispersions of graphene nanosheets [J]. Journal of Materials Chemistry, **2010**, 20 (21): 4328-4332.

[22] Cuong Tran Viet, Pham Viet Hung, Tran Quang Trung, et al. Photoluminescence and Raman studies of graphene thin films prepared by reduction of graphene oxide [J]. Materials Letters, **2010**, 64 (3): 399-401.

[23] Erickson Kris, Erni Rolf, Lee Zonghoon, et al. Determination of the local chemical structure of graphene oxide and reduced graphene oxide [J]. Advanced Materials, **2010**, 22 (40): 4467-4472.

[24] Shang Jingzhi, Lin Ma, Li Jiewei, et al. The Origin of Fluorescence from Graphene Oxide [J]. Sci Rep, **2012**, 2 (6108): 792.

[25] Cuong Tran Viet, Viet HungPham, Woo Shin Eun, et al. Temperature-dependent photoluminescence from chemically and thermally reduced graphene [J]. Applied Physics Letters, **2011**, 99 (4): 041905.

[26] Li Ming, Cushing Scott K., Zhou Xuejiao, et al. Fingerprinting photoluminescence of functional groups in graphene oxide [J]. Journal of Materials Chemistry, **2012**, 22 (44): 23374-23379.

[27] Kochmann Sven, Hirsch Thomas, Wolfbeis Otto. The pH Dependence of the Total Fluorescence of Graphite Oxide [J]. Journal of Fluorescence, **2012**, 22 (3): 849-855.

[28] Dong Yongqiang, Shao Jingwei, Chen Congqiang, et al. Blue luminescent graphene quantum dots and graphene oxide prepared by tuning the carbonization degree of citric acid [J]. Carbon, **2012**, 50 (12): 4738-4743.

[29] Brodie B. C. On the Atomic Weight of Graphite [J]. Philosophical Transactions of the Royal Society of London, **2009**, 149 (1): 249-259.

[30] Hummers W S, Offeman R E. Preparation of Graphitic Oxide [J]. Journal of the American Chemical Society, **1958**, 208: 1334-1339.

[31] Gokus Tobias, Raveendran-Nair Rahul, Bonetti A, et al. Making graphene luminescent by oxygen plasma treatment [J]. Acs Nano, **2009**, 3 (12): 3963-3968.

[32] Sun Xiaoming, Liu Zhuang, Welsher Kevin, et al. Nano-graphene oxide for cellular imaging and drug delivery [J]. Nano Research, **2008**, 1 (3): 203-212.

[33] Kovtyukhova Nina I., Ollivier Patricia J., Martin Benjamin R., et al. Layer-by-Layer Assembly of Ultrathin Composite Films from Micron-Sized Graphite Oxide Sheets and Polycations [J]. Chemistry of Materials, **1999**, 11 (3): 771-778.

[34] Dimiev Ayrat M, Eigler Siegfried. Graphene oxide: fundamentals and applications [M]. City: John Wiley & Sons, **2016**.

[35] Robertson J. π-bonded clusters in amorphous carbon materials [J]. Philosophical Magazine B, **1992**, 66 (2): 199-209.

[36] Schniepp Hannes, Li Je-Luen, J McAllister Michael, et al. Functionalized single graphene sheets derived from splitting graphite oxide [J]. Journal of Physical Chemistry B, **2006**, 110 (17): 8535-8539.

[37] Li Xiaolin, Zhang Guangyu, Bai Xuedong, et al. Highly conducting graphene sheets and Langmuir-Blodgett films [J]. Nature Nanotechnology, **2008**, 3 (9): 538-542.

[38] Hasan Md Tanvir, J Senger Brian, Mulford Price, et al. Modifying optical properties of reduced/graphene oxide with controlled ozone and thermal treatment in aqueous suspensions [J]. Nanotechnology, **2017**, 28 (6): 065705.

[39] Zhu Shoujun, Zhang Junhu, Qiao Chunyan, et al. Strongly green-photoluminescent graphene quantum dots for bioimaging applications [J]. Chemical Communications, **2011**, 47 (24): 6858-6860.

[40] Liu Yan, Liu Chun Yan, Liu Yun. Investigation on fluorescence quenching of dyes by graphite oxide and graphene [J]. Applied Surface Science, **2011**, 257 (13): 5513-5518.

[41] Povedailo V A, Ronishenko B V, Stepuro V I, et al. Fluorescence Quenching of Dyes by Graphene Oxide [J]. Journal of Applied Spectroscopy, **2018**, 85 (4): 605-610.

[42] Jaemyung Kim, Cote Laura J, Franklin Kim, et al. Visualizing graphene based sheets by fluorescence quenching microscopy [J]. Journal of the American Chemical Society, **2010**, 132 (1): 260-267.

[43] Jiang Xiao-Fang, Lakshminarayana Polavarapu Ting Neo, Shu, Venkatesan T, et al. Graphene Oxides as Tunable Broadband Nonlinear Optical Materials for Femtosecond Laser Pulses [J]. Journal of Physical Chemistry Letters, **2012**, 3 (6): 785-790.

[44] Mariserla Bala Murali Krishna, Nutalapati Venkatramaiah, Venkatesan R., et al. Synthesis and structural, spectroscopic and nonlinear optical measurements of graphene oxide and its composites with metal and metal free porphyrins [J]. Journal of Materials Chemistry, **2012**, 22 (7): 3059-3068.

[45] Wang Jun, Hernandez Yenny, Lotya Mustafa, et al. Broadband Nonlinear Optical Response of Graphene Dispersions [J]. Advanced Materials, **2010**, 21 (24): 2430-2435.

[46] Xiaorui Zheng, Baohua Jia, Xi Chen, et al. In situ third-order non-linear responses during laser reduction of graphene oxide thin films towards on-chip non-linear photonic devices [J]. Advanced Materials, **2014**,

26（17）：2699-2703.
[47] Liang Haifeng. Mid-infrared response of reduced graphene oxide and its high-temperature coefficient of resistance [J]. Aip Advances，**2014**，4（10）：107131.
[48] Chang Haixin，Zhenhua Sun，Saito Mitsuhiro，et al. Regulating infrared photoresponses in reduced graphene oxide phototransistors by defect and atomic structure control [J]. Acs Nano，**2013**，7（7）：6310-6320.
[49] Park Jiwoong，H Ahn Y，Ruiz-Vargas Carlos. Imaging of photocurrent generation and collection in single-layer graphene [J]. Nano Letters，**2009**，9（5）：1742-1746.
[50] Liu Chang-Hua，Chang You-Chia，B Norris Theodore，et al. Graphene photodetectors with ultra-broad-band and high responsivity at room temperature [J]. Nature Nanotechnology，**2014**，9（4）：273-278.
[51] Liu Jingjing，Yin Yanlong，Yu Longhai，et al. Silicon-graphene conductive photodetector with ultra-high responsivity [J]. Scientific Reports，**2017**，7：40904.
[52] Lee Eduardo J H，Kannan Balasubramanian，Ralf Thomas Weitz，et al. Contact and edge effects in graphene devices [J]. Nature Nanotechnology，**2008**，3（8）：486.
[53] Xia Fengnian，Mueller Thomas，Golizadeh-Mojarad Roksana，et al. Photocurrent imaging and efficient photon detection in a graphene transistor [J]. Nano Letters，**2009**，9（3）：1039-1044.
[54] Jabbarzadeh Farnaz，Siahsar Mehrdad，Dolatyari Mahboubeh，et al. Modification of graphene oxide for applying as mid-infrared photodetector [J]. Applied Physics B，**2015**，120（4）：637-643.
[55] Khan Mustaque Ali，Nanda Karuna Kar，Krupanidhi S B Reduced graphene oxide film based highly responsive infrared detector [J]. Materials Research Express，**2017**，4（8）：085603.
[56] Li Xinming，Zhu Hongwei，Wang Kunlin，et al. Graphene-on-silicon Schottky junction solar cells [J]. Advanced Materials，**2010**，22（25）：2743-2748.
[57] Sinha Dhiraj，Lee J. U. Ideal graphene/silicon Schottky junction diodes [J]. Nano Letters，**2014**，14（8）：4660.
[58] Chen Chun-Chung，Aykol Mehmet，Chang Chia-Chi，et al. Graphene-silicon Schottky diodes [J]. Nano Letters，**2011**，11（5）：1863-1867.
[59] Zeng Longhui，Wang Mingzheng，Hu Han，et al. Monolayer graphene/germanium Schottky junction as high-performance self-driven infrared light photodetector [J]. Acs Appl Mater Interfaces，**2013**，5（19）：9362-9366.
[60] Gao Zhiwei，Jin Weifeng，Zhou yu，et al. Self-powered flexible and transparent photovoltaic detectors based on CdSe nanobelt/graphene Schottky junctions [J]. Nanoscale，**2013**，5（12）：5576-5581.
[61] Ye Yu，Gan Lin，Dai Lun，et al. Multicolor graphene nanoribbon/semiconductor nanowire heterojunction light-emitting diodes [J]. Journal of Materials Chemistry，**2011**，21（32）：11760-11763.
[62] Jong Yu Woo，Li Zheng，Zhou Hailong，et al. Vertically stacked multi-heterostructures of layered materials for logic transistors and complementary inverters [J]. Nature Materials，**2013**，12（3）：246-252.
[63] An Xiaohong，Liu Fangze，Jung Yung，et al. Tunable graphene-silicon heterojunctions for ultrasensitive photodetection [J]. Nano Letters，**2013**，13（3）：909-916.
[64] Fernandes Gustavo，Kim Jin Ho，Oller Declan，et al. Reduced graphene oxide mid-infrared photodetector at 300 K [J]. Applied Physics Letters，**2015**，107（11）：111111.
[65] Cao Yang，Zhu Jiayi，Xu Jia，et al. Ultra-broadband photodetector for the visible to terahertz range by self-assembling reduced graphene oxide-silicon nanowire array heterojunctions [J]. Small，**2014**，10（12）：2345-2351.
[66] Singh Manjri，Kumar Gaurav，Prakash Nisha，et al. Large bandgap reduced graphene oxide（rGO）based

n-p+ heterojunction photodetector with improved NIR performance [J]. Semiconductor Science Technology, **2018**, 33 (4): 045012.

[67] Zhu Miao, Li X. M., Guo Yibo, et al. Vertical junction photodetectors based on reduced graphene oxide/silicon Schottky diodes [J]. Nanoscale, **2014**, 6 (9): 4909-4914.

[68] Guo Linjuan, Yang Zheng, Dou Xincun. Artificial Olfactory System for Trace Identification of Explosive Vapors Realized by Optoelectronic Schottky Sensing [J]. Advanced Materials, **2016**, 29 (5): 1604528.

[69] Dehkharghani Mehrdad Naemi, Rajabi Ali, Amirmazlaghani Mina. Responsivity improvement of Si-based NIR photodetector using reduced Graphene Oxide [C]. Electrical Engineering. 2017, 523-526.

[70] Yang Hua, Cao Yang, He Junhui, et al. Highly conductive free-standing reduced graphene oxide thin films for fast photoelectric devices [J]. Carbon, **2017**, 115: 561-570.

[71] Abid, Sehrawat Poonam, Islam S. S., et al. Reduced graphene oxide (rGO) based wideband optical sensor and the role of Temperature, Defect States and Quantum Efficiency [J]. Scientific Reports, **2018**, 8 (1): 3537.

[72] Bao Qiaoliang, Han Zhang, Yu Wang, et al. Atomic-Layer Graphene as a Saturable Absorber for Ultrafast Pulsed Lasers [J]. Advanced Functional Materials, **2010**, 19 (19): 3077-3083.

[73] Luo Zhengqian, Zhou Min, Weng Jian, et al. Graphene-based passively Q-switched dual-wavelength erbium-doped fiber laser [J]. Optics Letters, **2010**, 35 (21): 3709-11.

[74] Popa D., Sun Zicai, Hasan Tawfique, et al. Graphene Q-switched, tunable fiber laser [J]. Applied Physics Letters, **2011**, 98 (7): 073106.

[75] Wang Y G, Chen H, Hsieh W, et al. Mode-locked Nd: GdVO4 laser with graphene oxide/polyvinyl alcohol composite material absorber as well as an output coupler [J]. Optics Communications, **2013**, 289 (Complete): 119-122.

[76] Yu Zhenhua, Song Yanrong, Dong Xinzheng, et al. Watt-level passively Q-switched double-cladding fiber laser based on graphene oxide saturable absorber [J]. Applied Optics, **2013**, 52 (29): 7127-7131.

[77] Chang Jianhua, Li Hanhan, Yang Zhenbo, et al. Efficient and compact Q-switched green laser using graphene oxide as saturable absorber [J]. Optics Laser Technology, **2018**, 98: 134-138.

[78] Ryzhii Victor, Satou Akira, and Otsuji Taiichi. Plasma waves in two-dimensional electron-hole system in gated graphene heterostructures [J]. Journal of Applied Physics, **2007**, 101 (2): 0245091-0245095.

[79] Satou Akira, T. Vasko F, Ryzhii Victor. Nonequilibrium carriers in intrinsic graphene under interband photoexcitation [J]. Phys. rev. b, **2008**, 78 (11): 1884-1898.

[80] Ryzhii Victor, Ryzhii Maxim, Satou Akira, et al. Feasibility of terahertz lasing in optically pumped epitaxial multiple graphene layer structures [J]. Journal of Applied Physics, **2009**, 106 (8): 19912.

[81] Ryzhii Victor, Ryzhii Maxim, Otsuji Taiichi. Population inversion of photoexcited electrons and holes in graphene and its negative terahertz conductivity [J]. Physica Status Solidi, **2010**, 5 (1): 261-264.

[82] Karasawa Hiromi, Komori Tsuneyoshi, Watanabe Takayuki, et al. Observation of Amplified Stimulated Terahertz Emission from Optically Pumped Heteroepitaxial Graphene-on-Silicon Materials [J]. Journal of Infrared Millimeter Terahertz Waves, **2011**, 32 (5): 655-665.

[83] Ryzhii Victor, Ryzhii Maxim, Otsuji Taiichi. Tunneling Recombination in Optically Pumped Graphene with Electron-Hole Puddles [J]. Applied Physics Letters, **2011**, 99 (17): 109-R.

[84] Wang Huan, Zhou Yixuan, Yao Zehan, et al. Terahertz generation from reduced graphene oxide [J]. Carbon, **2018**.

[85] Xu Jia, Liu Jiang, Wu Sida, et al. Graphene oxide mode-locked femtosecond erbium-doped fiber lasers [J]. Optics Express, **2012**, 20 (14): 15474-15480.

[86] Sridevi S, Vasu K. S., Jayaraman N., et al. Optical bio-sensing devices based on etched fiber Bragg gratings coated with carbon nanotubes and graphene oxide along with a specific dendrimer [J]. Sensors Actuators B Chemical, **2014**, 195 (5): 150-155.

[87] Alberto Nélia, Vigário César, Duarte Daniel, et al. Characterization of Graphene Oxide Coatings onto Optical Fibers for Sensing Applications [J]. Materials Today Proceedings, **2015**, 2 (1): 171-177.

[88] Xiao Yi, Zhang Jun, Cai Xiang, et al. Reduced graphene oxide for fiber-optic humidity sensing [J]. Optics Express, **2014**, 22 (25): 31555-31567.

[89] Mishra S K, Tripathi S N, Choudhary V, et al. SPR based fibre optic ammonia gas sensor utilizing nanocomposite film of PMMA/reduced graphene oxide prepared by in situ polymerization [J]. Sensors and Actuators B: Chemical, **2014**, 199: 190-200.

[90] Yu Caibin, Wu Yu, Liu Xiaolei, et al. Miniature fiber-optic NH_3 gas sensor based on Pt nanoparticle-incorporated graphene oxide [J]. Sensors Actuators B Chemical, **2017**, 244: 107-113.